Hayk Sedrakyan
Nairi Sedrakyan

AMC 12 preparation book

American Mathematics Competitions
preparation book

2021

About RSM

According to Russian tradition - the study of mathematics is the preeminent tool of mental development, and of learning to think powerfully. The top academic minds of the Soviet Union were tasked with developing a right way - a curriculum and methodology that would lead generations of students to their full potential. The resulting methods and textbooks came to be **used by elite schools** globally including in China, India, and Europe.

In time, the Russian methodology gave rise to generations of thinkers with a deep mathematical foundation, who could think critically, creatively, logically, and who welcomed challenge and the unknown. The result was having one of the strongest math schools in the world, standing next to the worlds best math schools, such as French or German ones.

The Russian School of Mathematics (RSM) was opened by two such immigrant women, who disappointed by the level of math education in the United States, opened a school for their own children and the children of their community. The curriculum and methodology, perfected over 20 years by a team of gifted academics, inspired by elite mathematical schools of the former Soviet Union, **adapted for the American environment**.

Today, RSMs award-winning, after-school math enrichment program serves about 40,000 K-12 students across North America. Ranked among the **top 10** schools with our students some of the *brightest young people in the world* by the the John Hopkins University Center for Talented Youth, RSM helps students develop a **deep and life-long understanding of mathematics**, as well as an advanced way of thinking and approaching problems.

In 2017, in order to develop and create a top-level math competition preparation program and math competition curriculum for RSM, the authors have started a collaboration with the newly opened math competitions department of RSM, Boston, USA. Now, this department includes hundreds of strongest students and dozens of math competition coaches. The authors would like to thank RSM for the support during the publication of this book.

Russian School of Mathematics 2021 ©

About the authors

Hayk Sedrakyan is an IMO medal winner, professional mathematical Olympiad coach in greater Boston area, Massachusetts, USA. He is the Dean of math competition preparation department at RSM. He has been a Professor of mathematics in Paris and has a PhD in mathematics (optimal control and game theory) from the UPMC - Sorbonne University, Paris, France. Hayk is a Doctor of mathematical sciences in USA, France, Armenia and holds three master's degrees in mathematics from institutions in Germany, Austria, Armenia and has spent a small part of his PhD studies in Italy. Hayk Sedrakyan has worked as a scientific researcher for the European Commission (sadco project) and has been one of the Team Leaders at Harvard-MIT Mathematics Tournament (HMMT). He took part in the International Mathematical Olympiads (IMO) in United Kingdom, Japan and Greece. Hayk has been elected as the President of the students general assembly and a member of the management board of the *Cite Internationale Universitaire de Paris* (10,000 students, 162 different nationalities) and the same year they were nominated for the Nobel Peace Prize.

Nairi Sedrakyan is involved in national and international mathematical Olympiads having been the President of Armenian Mathematics Olympiads and a member of the IMO problem selection committee. He is the author of the most difficult problem ever proposed in the history of the International Mathematical Olympiad (IMO), 5th problem of 37th IMO. This problem is considered to be the hardest problems ever in the IMO because none of the members of the strongest teams (national Olympic teams of China, USA, Russia) succeeded to solve it correctly and because national Olympic team of China (the strongest team in the IMO) obtained a cumulative result equal to 0 points and was ranked 6th in the final ranking of the countries instead of the usual 1st or 2nd place. The British 2014 film X+Y, released in the USA as *A Brilliant Young Mind*, inspired by the film *Beautiful Young Minds* (focuses on an English mathematical genius chosen to represent the United Kingdom at the IMO) also states that this problem is the hardest problem ever proposed in the history of the IMO (minutes 9:40-10:30). Nairi Sedrakyan's students (including his son Hayk Sedrakyan) have received 20 medals in the International Mathematical Olympiad (IMO), including Gold and Silver medals.

Overview

This book consists **only of author-created problems with author-prepared solutions (never published before)** and it is intended as a teacher's manual of mathematics, a self-study handbook for high-school students and mathematical competitors interested in American Mathematics Competitions (especially AMC 12). The book teaches problem solving strategies and aids to improve problem solving skills. The book includes a list of the most useful theorems and formulas for AMC 12, it also includes 14 sets of author-created AMC 12 type practice tests (350 author-created AMC 12 type problems and their detailed solutions). National Math Competition Preparation (NMCP) program of RSM used part of these 14 sets of practice tests to train students for AMC 12, as a result **75 % of NMCP high school students qualified for AIME**.
The authors provide both a list of answers for all 14 sets of author-created AMC 12 type practice tests and author-prepared solutions for each problem.

Keywords: AMC 12 preparation book, American Mathematics Competitions preparation, AMC problems and solutions sample tests, AMC 12 problems and solutions sample tests, AMC 12 questions sample tests, AMC problem solving strategies.

In case of any comments please contact sedrakyan.hayk@gmail.com

Mathematical competition is not about winning or losing, it is about mastering the art of thinking creatively and smart.
The true effectiveness of math competition training is to develop creative thinking and problem solving skills that will help in future careers.
Hayk Sedrakyan.

Contents

1 List of the most useful formulas and theorems for AMC 12 — 12
  1.1 Algebra (AMC 12): the most useful formulas and theorems — 12
    1.1.1 Basic factorization formulas and binomial expansions. — 12
    1.1.2 Quadratic and cubic equations, Vieta's formulas. — 13
    1.1.3 Bézout's little theorem, linear factorization, Newton's divided difference formula. — 14
    1.1.4 Formulas for arithmetic, geometric and Fibonacci sequences — 15
    1.1.5 Logarithms and rules for logarithms. — 16
    1.1.6 HM-GM-AM-QM inequalities. — 17
    1.1.7 Cauchy-Bunyakovsky-Schwarz inequality. — 17
    1.1.8 Sedrakyan's inequality. — 17
    1.1.9 Sedrakyan's power sums triangle. — 18
    1.1.10 Cartesian coordinate system, some important formulas. — 19
  1.2 Geometry (AMC 12): the most useful formulas and theorems — 21
    1.2.1 Formulas for plane shapes. — 21
    1.2.2 Tangential, cyclic, bicentric and extangential quadrilaterals, Pitot's theorem. — 22
    1.2.3 Heron's formula and Brahmagupha's formula. — 23
    1.2.4 Ceva's theorem, Menelaus' theorem, Stewart's theorem, Ptolemy's theorem. — 24
    1.2.5 Bretschneider's formula, *diagonals and sides* area formula of a quadrilateral. — 24
    1.2.6 Parameshvara's formula for circumradius. — 25
    1.2.7 Formulas for volume and surface areas of three-dimensional shapes. — 26
    1.2.8 Trigonometric identities. — 27
    1.2.9 Complex numbers, de Moivre's formula, Euler's formula. — 29
  1.3 Number theory (AMC 12): the most useful formulas and theorems — 30
    1.3.1 Unique-prime-factorization theorem (fundamental theorem of arithmetic). — 30
    1.3.2 Number of divisors of a composite number, sum and product of divisors. — 30
    1.3.3 One useful lemma. — 30
    1.3.4 Modular arithmetic and congruence relation. — 31
    1.3.5 Fermat's little theorem and Wilson's theorem. — 31
  1.4 Combinatorics and probability (AMC 12): the most useful formulas and theorems — 32
    1.4.1 Rule of sum and rule of product. — 32
    1.4.2 Permutations. — 32
    1.4.3 Combinations. — 32
    1.4.4 Stars and bars technique (integer equations). — 33
    1.4.5 Probability. — 33

2 AMC 12 type practice tests — 35
  2.1 AMC 12 type practice test 1 — 35
  2.2 AMC 12 type practice test 2 — 39
  2.3 AMC 12 type practice test 3 — 42
  2.4 AMC 12 type practice test 4 — 46

2.5	AMC 12 type practice test 5	50
2.6	AMC 12 type practice test 6	53
2.7	AMC 12 type practice test 7	57
2.8	AMC 12 type practice test 8	61
2.9	AMC 12 type practice test 9	64
2.10	AMC 12 type practice test 10	67
2.11	AMC 12 type practice test 11	70
2.12	AMC 12 type practice test 12	73
2.13	AMC 12 type practice test 13	77
2.14	AMC 12 type practice test 14	80

3 Answers   83
3.1 Answers of AMC 12 type practice tests 83

4 Solutions   87
4.1	Solutions of AMC 12 type practice test 1	87
4.2	Solutions of AMC 12 type practice test 2	101
4.3	Solutions of AMC 12 type practice test 3	114
4.4	Solutions of AMC 12 type practice test 4	126
4.5	Solutions of AMC 12 type practice test 5	138
4.6	Solutions of AMC 12 type practice test 6	152
4.7	Solutions of AMC 12 type practice test 7	164
4.8	Solutions of AMC 12 type practice test 8	181
4.9	Solutions of AMC 12 type practice test 9	197
4.10	Solutions of AMC 12 type practice test 10	208
4.11	Solutions of AMC 12 type practice test 11	222
4.12	Solutions of AMC 12 type practice test 12	238
4.13	Solutions of AMC 12 type practice test 13	254
4.14	Solutions of AMC 12 type practice test 14	267

Acknowledgment

The authors would like to thank their family for the support.

Introduction

What is AMC 12?

The American Mathematics Competitions (AMC) are math competitions in middle and high school, organized by the Mathematical Association of America (MAA), that begin the multy-level selection process of the United States team for the International Mathematical Olympiad (IMO).
There are three levels: **AMC 8** (for students in grade 8 or below), **AMC 10** (for students in grade 10 or below), **AMC 12** (for students in grade 12 or below).
The track usually starts with AMC 8. It serves as a practice to prepare for AMC 10 and AMC 12.
Students who took part in AMC 10 and were in the top 2.5 percent are invited to take part in the American Invitational Mathematics Examination (AIME).
Students who took part in AMC 12 and were in the top 5 percent are also invited to take part in the American Invitational Mathematics Examination (AIME).
Students who qualify through AMC 10 to take part in AIME and perform well enough on AIME are then invited to the United States of America Junior Mathematical Olympiad (USAJMO).
Students who qualify through AMC 12 to take part in AIME and perform well enough on AIME are then invited to the United States of America Mathematical Olympiad (USAMO).
Qualifying for USAJMO or USAMO is widely considered as one of the most prestigious awards for high school students in the United States.
Top 30 performing students on USAJMO or USAMO are invited to go to the Mathematical Olympiad Summer Program (MOSP or MOP) with the goal of providing them a deep foundation in math Olympiads.
Top 12 performing students are invited to take the Team Selection Test (TST).
Top 6 performing students are selected from these top 12 students to form the United States International Math Olympiad team (US IMO team).

What benefits does AMC participation give to students?

There are many benefits for taking part and doing well on AMC. The list of top scoring students becomes available to colleges, institutions and programs interested in attracting students with strong math background. This gives the applicant an advantage over the other applicants during the admission process. Moreover, sometimes even the best US (or international) colleges take these achievements into consideration and offer a full study scholarship.
Besides this, top scorers on AMC 10 and AMC 12 qualify to participate in the next rounds of math competitions, leading to become a team member of the United States team to take part in the International Mathematical Olympiad (IMO). The most prestigous math Olympiad in the world.
One of the most important benefits is that mathematical competitions are not about winning or losing, they are about mastering the art of thinking creatively and smart. The true effectiveness of math competition training is to develop creative thinking and problem solving skills that will help in future careers.

Strategy advices: becoming more strategic

Strategic decision 1. AMC 12 is a math competition where the problems change anually, but the list of possible topics does not change. Taking this into consideration, the authors provide in the book a list of the most useful formulas and theorems for AMC 12. Mastering this list of formulas and theorems plays a crucial role for student's performance during the actual AMC 12.

Strategic decision 2. The authors aimed to create AMC 12 type practice tests in order to help students to get prepared for AMC 12. **All problems in the book are author-created problems with author-prepared solutions (never published before)**. They made the tests as close as possible to the topics of AMC 12, but intentionally made most of the tests *slightly* harder than actual AMC 12 tests, believing that if students train to solve slightly harder problems, then it will be easier for them to solve AMC 12 problems during the actual competition. As AMC 12 is 75 minutes math competition with 25 problems of increasing difficulty, this type of preparation also helps to save some time during the actual AMC 12 competition and avoid the possibility of running out of the time.

Strategic decision 3. The best strategy for AMC 12 is to solve all 25 problems correctly within given 75 minutes. This is not always the case and sometimes students need to follow certain strategy to get qualified to AIME (American Invitational Mathematics Examination). With the strategy described in this paragraph, getting qualified to AIME through AMC 12 becomes easier than getting qualified to AIME through AMC 10, even for a ninth grader. AIME cutoff score differs from year to year, as those who rank in the top 5% on AMC 12 get qualified to AIME and those who rank in the top 2.5% on AMC 10 get qualified to AIME. If student's main goal is to get qualified to AIME, then one needs to understand AMC 12 and AMC 10 grading systems. Both are multiple choice math competition tests, where correct answers are worth 6 points, incorrect answers are worth 0 points and unanswered questions are worth 1.5 points. From year 2000 to year 2020 AIME cutoff scores (through AMC 12) were from 90 to 100, and (through AMC 10) were from 100 to 120. This means that if a student manages to solve at least 17 problems correctly in AMC 12, then the student gets qualified to AIME, as $17 \cdot 6 = 102$. Scoring at least 120 in AMC 10, means solving correctly about 20 problems. Solving at least 17 problems correctly may be challenging, often students run out of time and try to guess the answers. Students who realize that they are going to run out of time, their strategy to get qualified to AIME can be using the advantages of the scoring system of AMC 12, for example they can concentrate on the easiest 14 problems of the test, solve them correctly and do not answer the other 11 problems. So, they have 75 minutes to solve the easiest 14 problems and they will get $11 \cdot 1.5 = 16.5$ points for skipping 11 problems. Their total score will be $14 \cdot 6 + 11 \cdot 1.5 = 100.5$ and they can get qualified to AIME. Note that every year about 10 problems are exactly the same in AMC 10 and AMC 12 tests. This, means that if the student uses this strategy getting qualified to AIME through AMC 12 gets easier than through AMC 10, as talking roughtly during AMC 12 student needs to solve 14 problems of the same difficulty as in AMC 10, but AIME cutoff through AMC 12 is lower. We do **not** encourage this strategy, as long as the **only** goal of the student is to get qualified to AIME and not to go further to USAMO.

Strategic decision 4. USAMO qualificaition is determined based on AMC 12 score plus AIME score. In the previous paragraph, we have already explained AMC 12 grading system. AIME is 3 hours math competition with 15 problems of increasing difficulty, such that each answer is an integer number between 0 and 999 (both 0 and 999 inclusive). Each correct answer is scored as one point, each incorrect or blank answer is scored as zero point. Therefore, AIME final score is an integer number from 0 to 15 (both 0 and 15 inclusive). To determine eligibility for the USAMO student's score on AMC 12 is added to 10 times the score on AIME. For example, if a student scores 120 on AMC 12 and manages to solve 10 problems correctly on AIME, then student's final score is $120 + 10 \cdot 10 = 220$. The cutoff for getting qualified to USAMO is usually from 220 to 230 combined points. Taking this into consideration, if your goal is not only to get qualified to USAMO, but also to go further to IMO, then we encourage you not to use the strategy described in the previous paragraph. In this case, simply try to do your best both in AMC 12 and in AIME in order to score as high as possible (the most diserable is at least 230 combined points).

Strategic decision 5. Do not worry and **keep in mind that mathematical competitions are not about winning or losing, they are about mastering the art of thinking creatively and smart**. So, no matter whether the student gets qualified to AIME or not, with proper math competition training (in the end) the student always wins. Good luck.

Chapter 1

List of the most useful formulas and theorems for AMC 12

As a reference the authors would like to provide a list of the most useful formulas and theorems for AMC 12. This list is very useful and important not only for AMC 12, but also for AIME, USAMO, IMO and various mathematical competitions. Taking into consideration that the main topics of AMC 12 are algebra, geometry, number theory, combinatorics and probability, we will divide this chapter into 4 sections, respectively.

1.1 Algebra (AMC 12): the most useful formulas and theorems

1.1.1 Basic factorization formulas and binomial expansions.

$a^2 - b^2 = (a-b)(a+b)$.
$(a-b)^2 = a^2 - 2ab + b^2$.
$a^2 + b^2 = (a-b)^2 + 2ab$.
$(a+b)^2 = a^2 + 2ab + b^2$.
$a^2 + b^2 = (a+b)^2 - 2ab$.
$(a^2+b^2)(c^2+d^2) = (ac+bd)^2 + (ad-bc)^2$.
$(a^2+b^2)(c^2+d^2) = (ac-bd)^2 + (ad+bc)^2$.
$(a+b+c)^2 = a^2 + b^2 + c^2 + 2(ab+bc+ac)$.

$a^3 - b^3 = (a-b)(a^2+ab+b^2)$.
$a^3 + b^3 = (a+b)(a^2-ab+b^2)$.
$(a-b)^3 = a^3 - 3a^2b + 3ab^2 - b^3$.
$a^3 - b^3 = (a-b)^3 - 3ab(a-b)$.
$(a+b)^3 = a^3 + 3a^2b + 3ab^2 + b^3$.
$a^3 + b^3 = (a+b)^3 - 3ab(a+b)$.
$a^3 + b^3 + c^3 - 3abc = (a+b+c)(a^2+b^2+c^2-ab-bc-ac)$.
$a^3 + b^3 + c^3 - 3abc = \frac{1}{2}(a+b+c)\left((a-b)^2 + (b-c)^2 + (c-a)^2\right)$.
$(a+b+c)^3 = a^3 + b^3 + c^3 + 3(a+b)(b+c)(c+a)$.

$ab + a + b + 1 = (a+1)(b+1)$, mostly used in algebra and number theory problems where a and b are integers (for example, when we deal with divisibility problems).
$ab - a - b + 1 = (a-1)(b-1)$, mostly used in algebra and number theory problems where a and b are integers (for example, when we deal with divisibility problems).

1.1.2 Quadratic and cubic equations, Vieta's formulas.

Given a quadratic equation of a general form $ax^2 + bx + c = 0$, where x is an unknown, a, b, c are given coefficients and $a \neq 0$. Note that, any quadratic equation has at most two solutions. If there is no real solution, there are two complex solutions. If there are two coinciding solutions, then it is called a *double root*. $D = b^2 - 4ac$ is called *discriminant*.
If the discriminant is 0, then quadratic equation has one real solution.
If the discriminant is negative, then quadratic equation has no real solutions.
If the discriminant is positive, then quadratic equation has two real solutions.
These two real solutions x_1, x_2 are given by the following formulas:

$$x_1 = \frac{-b - \sqrt{b^2 - 4ac}}{2a},$$

$$x_2 = \frac{-b + \sqrt{b^2 - 4ac}}{2a}.$$

Factoring quadratic expressions. Note that in this case quadratic equation can be factored in the following way:
$$ax^2 + bx + c = a(x - x_1)(x - x_2).$$

Completing the square method. Note that, using the formula $A^2 + 2AB + B^2 = (A+B)^2$, we obtain

$$ax^2 + bx + c = a\left(x^2 + \frac{b}{a}x + \frac{c}{a}\right) = a\left(x^2 + \frac{b}{a}x + \left(\frac{b}{2a}\right)^2 - \left(\frac{b}{2a}\right)^2 + \frac{c}{a}\right) = a\left(\left(x + \frac{b}{2a}\right)^2 - \frac{b^2 - 4ac}{4a^2}\right).$$

Thus, it follows that
$$ax^2 + bx + c = a\left(x + \frac{b}{2a}\right)^2 - \frac{b^2 - 4ac}{4a}.$$

This method is called *completing the square* and it can be used for different purposes. For example, for finding the minimum and maximum value of a quadratic equation.
Minimum and maximum value of a quadratic expression. Minimum and maximum value of a quadratic expression $f(x) = ax^2 + bx + c, a \neq 0$ can be found in different ways, for example by completing the square, graphycally, using derivatives. Let us consider the following two cases:
Case 1. If $a > 0$, then the maximum value is infinite and from completing of the square form we obviously see that the minimum value occurs at $x = -\frac{b}{2a}$, (as any quadratic equation graphically represents a parabola, this point is also called the vertex of corresponding parabola open upward).
Note that in this case the minimum value of the quadratic equation is equal to:

$$f\left(-\frac{b}{2a}\right) = \frac{4ac - b^2}{4a}.$$

Case 2. If $a < 0$, then the minimum value is infinite and from completing of the square form we obviously see that the maximum value occurs at $x = -\frac{b}{2a}$, (at the vertex of the corresponding parabola open downward).
Note that in this case the maximum value of the quadratic equation is equal to:

$$f\left(-\frac{b}{2a}\right) = \frac{4ac - b^2}{4a}.$$

Vieta's formulas are named after a prominent French mathematician François Viète. We provide Vieta's formulas only for quadratic and cubic equations, as they are used widely in different math competitions.

Note that Vieta's formula holds true for a polynomial of any finite power.
Vieta's formula for a quadratic equation of a general form $ax^2 + bx + c = 0$, where $a \neq 0$, states that the sum of the roots (solutions) x_1, x_2 is equal to $-\dfrac{b}{a}$ and the product of the roots is equal to $\dfrac{c}{a}$. This can be rewritten in the following way:

$$ax^2 + bx + c = 0, a \neq 0,$$

$$\begin{cases} x_1 + x_2 = -\dfrac{b}{a}, \\ x_1 \cdot x_2 = \dfrac{c}{a}. \end{cases}$$

In the case of the cubic equation, we have that

$$ax^3 + bx^2 + cx + d = 0, a \neq 0,$$

$$\begin{cases} x_1 + x_2 + x_3 = -\dfrac{b}{a}, \\ x_1 \cdot x_2 + x_2 \cdot x_3 + x_1 \cdot x_3 = \dfrac{c}{a}, \\ x_1 \cdot x_2 \cdot x_3 = -\dfrac{d}{a}. \end{cases}$$

1.1.3 Bézout's little theorem, linear factorization, Newton's divided difference formula.

Bézout's little theorem, also called the *polinomial remainder theorem* is named after a French matematician *Étienne Bézout*. Before giving the formulation of the theorem, let us give the defition of *Euclidean division of polynomials*, also called *long division of polynomials*.

Euclidean division of polynomials (long division of polynomials). Let P and D be two polynomials, where D is not zero, then there exist polynomials Q (called a *quotient*) and R (called a *remainder*), such that

$$P = D \cdot Q + R,$$

where either $R = 0$ or the degree of polynomial R is less than the degree of polynomial D (denoted as $\deg(R) < \deg(D)$). Here P is called *the dividend* and D is called *the divisor*. Moreover, Q and R are uniquely defined.

Bézout's little theorem. The remainder of the division of a polynomial $P(x)$ by a linear polynomial $x - k$ is equal to $P(k)$.

This theorem is named after a French mathematician *Étienne Bézout*.

Factor theorem. A polynomial $P(x)$ has a factor $(x - k)$ if and only if $P(k) = 0$ (k is a root of polynomial $P(x)$).

Linear factorization theorem. Let n be a nonnegative integer and

$$P(x) = a_n x^n + a_{n-1} x^{n-1} + ... + a_1 x + a_0,$$

be a polynomial of degree n in variable x, where $a_n, a_{n-1}, ..., a_1, a_0$ are the coefficients of the polynomial and $a_n \neq 0$ (a_n is called the leading coefficient and a_0 is called the constant term). Then $P(x)$ is possible to represent as the product of n linear factors

$$P(x) = a_n(x - x_1)(x - x_2) \cdot ... \cdot (x - x_n),$$

where $x_1, x_2, ..., x_n$ are the roots of polynomial $P(x)$.

Remark. Note that some of the values of x_i can be complex, where $i = 1, 2, ..., n$.

Lemma (consequence of Newton's divided difference formula). Let $P(x)$ be a polynomial of degree n in variable x. Let $t_1, t_2, ..., t_n$ be any numbers, then there exist unique numbers $c_0, c_1, ..., c_n$, such that

$$P(x) = c_0 + c_1(x - t_1) + c_2(x - t_1)(x - t_2) + ... + c_n(x - t_1)(x - t_2)...(x - t_n).$$

Note that we modified and wrote in a simpler way the classical formulation of *Newton's divided difference formula* to be able to apply it to AMC problems (if needed). It is named after a prominent English mathematician, physicist and astronomer *Isaac Newton*.

Remark. Stating the values of numbers $c_0, c_1, ..., c_n$ is outside of the scope of this book, as their values are not used in AMC problems, only this representation is sometimes used in AMC problems. Nevertheless, the classical formulation of Newton's divided difference formula used in the literature provides formulas for the values of numbers $c_0, c_1, ..., c_n$.

1.1.4 Formulas for arithmetic, geometric and Fibonacci sequences

Definition (arithmetic sequence). A sequence of numbers such that the difference between the consecutive terms is constant number is called an *arithmetic sequence*.

This constant number is called the common difference and usually is denoted by d. For a positive integer n we denote n–th term of this sequence by a_n. Therefore, this definition means that

$$a_{n+1} = a_n + d,$$

where $n = 1, 2, ...$

The sum of the first n terms is usually denote by S_n, that is $S_n = a_1 + a_2 + ... + a_n$. We provide the following two important formulas related to arithmetic sequences.

$$a_n = a_1 + (n-1)d,$$

and

$$S_n = \frac{a_1 + a_n}{2} \cdot n.$$

Definition (geometric sequence). A sequence of numbers where each term after the first is found by multiplying the previous one by a non-zero constant number is called a *geometric sequence*.

This constant number is called *common ratio* and usually is denoted by r. For a positive integer n we denote n–th term of this sequence by g_n. Therefore, this definition means that

$$g_{n+1} = g_n \cdot r,$$

where $n = 1, 2, ...$

The sum of the first n terms is usually denote by S_n, that is $S_n = g_1 + g_2 + ... + g_n$. We provide the following two important formulas related to geometric sequences.

$$g_{n+1} = g_1 \cdot r^n,$$

and

$$S_n = \frac{g_1(1 - r^n)}{1 - r},$$

where $r \neq 1$.
If $r = 1$, then

$$S_n = n \cdot g_1.$$

Infinite geometric sequences. The sum of all terms of an infinitie geometric sequence is denoted by
$$S_\infty = g_1 + g_2 + ... + g_n + ...,$$
and we take $-1 < r < 1$. We provide the following important formula for the sum of an infinite geometric sequence, where $-1 < r < 1$.
$$S_\infty = \frac{g_1}{1-r}.$$

Definition (Fibonacci sequence). A sequence of numbers where each number is the sum of the two preceding ones, starting from 0 and 1, is called the *Fibonacci sequence*.
It is named so after an italian mathematician *Leonardo Bonacci* also known as *Fibonacci*. The terms of this sequence are called *Fibonacci numbers* and are denoted by F_n, for a nonnegative integer n. According to the definition, we have that $F_0 = 0$, $F_1 = 1$ and for $n \geq 2$ it follows that
$$F_n = F_{n-1} + F_{n-2}.$$

Let us provide first few terms of the Fibonacci sequence: 0, 1, 1, 2, 3, 5, 8, 13, 21, 34, 55, 89, ...
Remark. In the literature, sometimes you can see that F_0 is ommited and the sequence starts with $F_1 = F_2 = 1$.

1.1.5 Logarithms and rules for logarithms.

Logarithms are another way of thinking about *exponents* and *exponential equations*. For example, given that 4 raised to 3rd power is equal to 64. This can be expressed by the following exponential equation $4^3 = 64$. In order to understand what is a logarithm, let us reformulate this question in the following way: 4 raised to which power is equal to 64? Obviously, the answer is 3. Introducing and using the notation of a logarithm, this can be rewritten as $\log_4 64 = 3$ and we read it as *log base four of sixty-four is three*.
Definition of logarithm. Let a, b be positive numbers and $b \neq 1$. We denote $x = \log_b a$, if and only if $b^x = a$.
Remark. Note that b is called the *base*, a is called the *argument*, x is called the *exponent*.
In the literature, the below mentioned two notations for logarithms are widely used. Taking this into consideration, we would like to introduce the following two notations to our readers.
Natural logarithm and common logarithm. A logarithm to the base of the mathematical constant e is called the **natural logarithm** of a positive number a and is denoted by $\ln a$, where $e \approx 2.718$ is *Euler's number* named after prominent Swiss mathematician *Leonhard Euler*. A logarithm to the base of 10 is called the **common logarithm** of a positive number a and is denoted by $\log a$.
Now, let us provide some properties and rules for logarithms, where a, b, c are positive numbers and $b \neq 1$.
Logarithm zero rule. $\log_b 1 = 0$.
Logarithm identity rule. $\log_b b = 1$.
Logarithm product rule. $\log_b(a \cdot c) = \log_b a + \log_b c$.
Logarithm quotient rule. $\log_b \left(\dfrac{a}{c}\right) = \log_b a - \log_b c$.
Logarithm power rule. $\log_b a^n = n \cdot \log_b a$.
Logarithm root rule. $\log_b \sqrt[n]{a} = \dfrac{1}{n} \cdot \log_b a$, where n is a positive integer greater than 1.
Logarithm change of base rule. $\log_b a = \dfrac{\log_c a}{\log_c b}$, where $c \neq 1$.
Logarithm base switch rule. $\log_b a = \dfrac{1}{\log_a b}$.
Logarithm power of base rule. $\log_{b^k} a = \dfrac{1}{k} \cdot \log_b a$, where k is a real number and $k \neq 0$.

1.1.6 HM-GM-AM-QM inequalities.

The following inequalities are called the HM (harmonic mean)-GM (geometric mean)-AM (arithmetic mean)-QM (quadratic mean) inequalities. At first, let us formulate the HM-GM-AM-QM inequalities for two positive real numbers a and b, that is

$$\frac{2}{\frac{1}{a}+\frac{1}{b}} \leq \sqrt{ab} \leq \frac{a+b}{2} \leq \sqrt{\frac{a^2+b^2}{2}}.$$

Note that the equality holds true if and only if $a = b$.
Now, let us formulate the HM-GM-AM-QM inequalities for any n positive real numbers $a_1, a_2, ..., a_n$, that is

$$\frac{n}{\frac{1}{a_1}+\frac{1}{a_2}+...+\frac{1}{a_n}} \leq \sqrt[n]{a_1 \cdot a_2 \cdot ... \cdot a_n} \leq \frac{a_1+a_2+...+a_n}{n} \leq \sqrt{\frac{a_1^2+a_2^2+...+a_n^2}{n}}.$$

Note that the equality holds true if and only if $a_1 = a_2 = ... = a_n$.

1.1.7 Cauchy-Bunyakovsky-Schwarz inequality.

Cauchy-Bunyakovsky-Schwarz inequality is considered to be one of the most important inequalities in mathematics. It is named after a prominent French mathematician *Augustin-Louis Cauchy*, a Russian mathematician *Viktor Bunyakovsky* and a Prussian mathematician *Karl Hermann Amandus Schwarz*.
Cauchy-Bunyakovsky-Schwarz inequality. For any real numbers $a_1, a_2, ..., a_n, b_1, b_2, ..., b_n$, we have

$$(a_1^2 + ... + a_n^2)(b_1^2 + ... + b_n^2) \geq (a_1 b_1 + ... + a_n b_n)^2.$$

1.1.8 Sedrakyan's inequality.

The following inequality is known as *Sedrakyan's inequality* (published in 1997), it is also called *Engel's form* (published in 1998) or *Titu's lemma* (published in 2003).
Sedrakyan's inequality. For any real numbers $a_1, a_2, ..., a_n$ and positive real numbers $b_1, b_2, ..., b_n$, we have

$$\frac{a_1^2}{b_1} + \frac{a_2^2}{b_2} + ... + \frac{a_n^2}{b_n} \geq \frac{(a_1+a_2+...+a_n)^2}{b_1+b_2+...+b_n}.$$

Moreover, the equality holds true when $\frac{a_1}{b_1} = \frac{a_2}{b_2} = ... = \frac{a_n}{b_n}$.
This inequality helps to prove very hard fractional inequalities in a straightforward and relatively simple way. It can be also used to solve fractional equations. Interested reader can find different generalizations of this inequality in the book *Algebraic Inequalities* of *Hayk Sedrakyan* and *Nairi Sedrakyan* (published by Springer).
Example: Application of Sedrakyan's inequality (IMO 1995, Problem 2). Let a, b, c be positive real numbers such that $abc = 1$. Prove that

$$\frac{1}{a^3(b+c)} + \frac{1}{b^3(c+a)} + \frac{1}{c^3(a+b)} \geq \frac{3}{2}.$$

Proof. Using Sedrakyan's inequality we are able to provide *one-line proof* for this IMO problem. Note that

$$\frac{\left(\frac{1}{a}\right)^2}{a(b+c)} + \frac{\left(\frac{1}{b}\right)^2}{b(c+a)} + \frac{\left(\frac{1}{c}\right)^2}{c(a+b)} \geq \frac{\left(\frac{1}{a}+\frac{1}{b}+\frac{1}{c}\right)^2}{2(ab+bc+ac)} = \frac{ab+bc+ac}{2} \geq \frac{3\sqrt[3]{a^2b^2c^2}}{2} = \frac{3}{2}.$$

1.1.9 Sedrakyan's power sums triangle.

In mathematical competitions we often deal with sums of the forms $1^k + 2^k + ... + n^k$, where n and k are positive integers. For $k = 1$ almost everyone remembers the formula $1 + 2 + ... + n = \dfrac{n(n+1)}{2}$, but very often in order to solve algebra or number theory problems students need to use the corresponding formulas for $1^2 + 2^2 + ... + n^2$, $1^3 + 2^3 + ... + n^3$, $1^4 + 2^4 + ... + n^4$ or higher powers. *Hayk Sedrakyan's* idea was to create a simple and self-constructive *Pascal-type triangle* for sums of powers.

Sedrakyan's power sums triangle. At first, let us provide the triangle and afterward explain how the self-constructive principle of this triangle works. This work was published in the book *Algebraic Inequalities* of *Hayk Sedrakyan* and *Nairi Sedrakyan* (published by Springer).

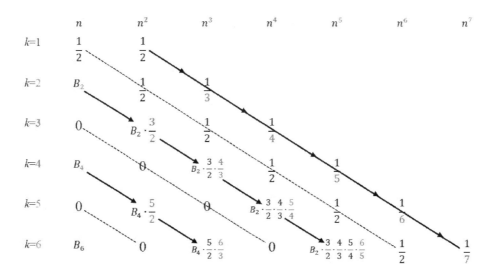

Self-construction principle. Almost all students remember the following well-known formula $1 + 2 + ... + n = \dfrac{n(n+1)}{2}$, we rewrite this formula as $1 + 2 + ... + n = \dfrac{1}{2} \cdot n + \dfrac{1}{2} \cdot n^2$. The first line of the considered *triangle* consists of these two coefficients $\left(\dfrac{1}{2}, \dfrac{1}{2}\right)$ and as in the case of *Pascal's triangle* using the coefficients of the first line we construct the next lines in order to find the value of the sum $1^k + 2^k + ... + n^k$ for $k = 2, 3, ...$ For this reason let us note the following:

1. The coefficients on the *dashed lines* remain constant.
2. The denominator of each coefficient on the *solid line* with arrows for each next row increases by 1. In other words, this means that this coefficient is equal to $\dfrac{1}{k+1}$.
3. *Small line segments with arrows* show how each next row can be obtained using the coefficients of the previous row. In other words, each time we multiply the coefficient of the previous row by a fraction with nominator equal to the value of k written in that row and with the denominator equal to the power of n written in that column.
4. If k is even, for example $k = 2$, then that row we start with B_2 and so on for the other even values of k, where $B_2, B_4, B_6...$ are *Bernoulli numbers*. The list of Bernoulli numbers is available in the literature and online, we would like to provide few of them $B_2 = \dfrac{1}{6}, B_4 = \dfrac{1}{-30}, B_6 = \dfrac{1}{42}, B_8 = \dfrac{1}{-30}, B_{10} = \dfrac{5}{66}, ...$

Example 1. Calculate the sum $1^3 + 2^3 + ... + n^3$.

Using this *power sums triangle* we deduce that

$$1^3 + 2^3 + ... + n^3 = 0 \cdot n + B_2 \cdot \dfrac{3}{2} \cdot n^2 + \dfrac{1}{2} \cdot n^3 + \dfrac{1}{4} \cdot n^4.$$

Hence, we obtain that
$$1^3 + 2^3 + \ldots + n^3 = \frac{1}{6} \cdot \frac{3}{2} \cdot n^2 + \frac{1}{2} \cdot n^3 + \frac{1}{4} \cdot n^4.$$

Thus, it follows that
$$1^3 + 2^3 + \ldots + n^3 = \frac{n^2}{4} + \frac{n^3}{2} + \frac{n^4}{4} = \left(\frac{n(n+1)}{2}\right)^2.$$

Example 2. Calculate the sum $1^6 + 2^6 + \ldots + n^6$.
Using this *power sums triangle* we deduce that
$$1^6 + 2^6 + \ldots + n^6 = B_6 \cdot n + 0 \cdot n^2 + B_4 \cdot \frac{5}{2} \cdot \frac{6}{3} \cdot n^3 + 0 \cdot n^4 + B_2 \cdot \frac{3}{2} \cdot \frac{4}{3} \cdot \frac{5}{4} \cdot \frac{6}{5} \cdot n^5 + \frac{1}{2} \cdot n^6 + \frac{1}{7} \cdot n^7.$$

Hence, we have that
$$1^6 + 2^6 + \ldots + n^6 = B_6 \cdot n + B_4 \cdot \frac{5}{2} \cdot \frac{6}{3} \cdot n^3 + B_2 \cdot \frac{6}{2} \cdot n^5 + \frac{1}{2} \cdot n^6 + \frac{1}{7} \cdot n^7.$$

Thus, it follows that
$$1^6 + 2^6 + \ldots + n^6 = \frac{1}{42} \cdot n - \frac{1}{30} \cdot \frac{5}{2} \cdot \frac{6}{3} \cdot n^3 + \frac{1}{6} \cdot \frac{6}{2} \cdot n^5 + \frac{1}{2} \cdot n^6 + \frac{1}{7} \cdot n^7.$$

Therefore, we obtain that
$$1^6 + 2^6 + \ldots + n^6 = \frac{n}{42} - \frac{n^3}{6} + \frac{n^5}{2} + \frac{n^6}{2} + \frac{n^7}{7}.$$

Remark. Note that in the literature a corresponding formula for the sum of 6th powers can be found, but for students it is almost impossible to memorize these formulas for the sum of any power, including the formula for the sum of 6th power. A straightforward verification shows that, the expression we have obtained is the same as the below mentioned formula that can be found in the literature, nevertheless our approach (*power sums triangle*) is self-constructive and can be easily applied to obtain a formula for the sum of any power.
$$1^6 + 2^6 + \ldots + n^6 = \frac{1}{42}n(n+1)(2n+1)(3n^4 + 6n^3 - 3n + 1).$$

1.1.10 Cartesian coordinate system, some important formulas.

Cartesian coordinate system. Cartesian coordinate system specifies each point uniquely in a plane by a set of numerical coordinates. Cartesian coordinate system was invented in 17th century by a prominent French mathematician *René Descartes*, his latinized name was *Cartesius*. His work revolutionized mathematics by providing link between *geometry* and *algebra*.

The distance formula in one dimension. The distance between two points $P_1 = x_1$ and $P_2 = x_2$ lying on the real number line can be found using the following formula.
$$d(P_1, P_2) = |x_1 - x_2|.$$

The distance formula in two dimensions. The distance between two points $P_1 = (x_1, y_1)$ and $P_2 = (x_2, y_2)$ of two dimensional xy–plane can be found using the following formula.
$$d(P_1, P_2) = \sqrt{(x_1 - x_2)^2 + (y_1 - y_2)^2}.$$

The distance formula in three dimensions. The distance between two points $P_1 = (x_1, y_1, z_1)$ and $P_2 = (x_2, y_2, z_2)$ of three dimensional xyz–space can be found using the following formula.
$$d(P_1, P_2) = \sqrt{(x_1 - x_2)^2 + (y_1 - y_2)^2 + (z_1 - z_2)^2}.$$

The distance formula in n−dimensional space. The distance between two points $P = (p_1, p_2, ..., p_n)$ and $Q = (q_1, q_2, ..., q_n)$ of n−dimensional space can be found using the following formula.
$$d(P,Q) = \sqrt{(q_1 - p_1)^2 + (q_2 - p_2)^2 + ... + (q_n - p_n)^2}.$$

The midpoint formula in n−dimensional space. The midpoint of a line segment in n−dimensional space with endpoints $P_1 = (x_1, x_2, ..., x_n)$ and $P_2 = (y_1, y_2, ..., y_n)$ is given by $\dfrac{P_1 + P_2}{2}$. In other words, the ith coordinate of the midpoint is $\dfrac{x_i + y_i}{2}$, where $i = 1, 2, ..., n$.

Equation of the line passing through two points. The equation of the line passing through two points $P_1 = (x_1, y_1)$ and $P_2 = (x_2, y_2)$ of two dimensional xy−plane can be found using the following formula.
$$y - y_1 = \frac{y_2 - y_1}{x_2 - x_1} \cdot (x - x_1).$$

Remark. The coefficient $m = \dfrac{y_2 - y_1}{x_2 - x_1}$ is called the slope of a staight line and represents the ratio of the "vertical change" to the "horizontal change".

Parallel lines. Two lines are parallel if they have the same slope.
That is, if the equation of the first line is $y = m_1 x + b_1$ and the equation of the second line is $y = m_2 x + b_2$, then these two lines are parallel if $m_1 = m_2$ and $b_1 \neq b_2$.

Perpendicular lines. Two lines are perpendicular, if the product of their slopes is equal to -1.
That is, if the equation of the first line is $y = m_1 x + b_1$ and the equation of the second line is $y = m_2 x + b_2$, then these two lines are perpendicular if $m_1 \cdot m_2 = -1$.

1.2 Geometry (AMC 12): the most useful formulas and theorems

1.2.1 Formulas for plane shapes.

Formula of the inradius of a right triangle. For inradius r of a right triangle with legs a, b and hypotenuse c, we have that
$$r = \frac{a+b-c}{2}.$$

Pythagorean theorem. For a right triangle with legs a, b and hypotenuse c, we have that
$$a^2 + b^2 = c^2.$$

Law of cosines. In a triangle ABC let a, b, c be the lengths of sides opposite to $\angle A, \angle B, \angle C$, respectively, then
$$\begin{cases} a^2 = b^2 + c^2 - 2bc \cdot \cos \angle A, \\ b^2 = a^2 + c^2 - 2ac \cdot \cos \angle B, \\ c^2 = a^2 + b^2 - 2ab \cdot \cos \angle C. \end{cases}$$

Remark. Note that the Pythagorean theorem is a special case of the law of cosines, because when $\angle C = 90°$ then $\cos \angle C = \cos 90° = 0$.

Law of sines. In a triangle ABC let a, b, c be the lengths of sides opposite to the angles A, B, C, respectively, then
$$\frac{a}{\sin A} = \frac{b}{\sin B} = \frac{c}{\sin C}.$$

Area of a triangle.
$$S = \frac{a \cdot h_a}{2} = \frac{rp}{2} = \frac{1}{2} ab \cdot \sin \gamma = \frac{abc}{4R},$$
where S is the area of the triangle, a, b, c are the side lengths, h_a is the length of the altitude to side a, p is the perimeter, r is the inradius, R is the circumradius, γ is the angle between the sides a and b.

Area of an equilateral triangle.
$$S = \frac{\sqrt{3}}{4} a^2,$$
where a is the side length.

Area of a rectangle and a square. Area of a rectangle is given by the following formula $S = a \cdot b$, where a, b are the side lengths of the rectangle. Note that in the case of the square $a = b$, therefore area of a sqaure is given by the following formula $S = a^2$.

Area of a trapezoid.
$$S = \frac{(b_1 + b_2) \cdot h}{2} = m \cdot h,$$
where b_1 and b_2 are the heights of the trapezoid, h is the height, m is the midsegment of the trapezoid (midsegment or midline is the segment connecting the midpoints of the two non-parallel sides).

Area of a quadrilateral (general formula). In the subsection *Bretschneider's formula, diagonals and sides area formula of a quadrilateral* we provide useful formulas to find the area of a quadrilateral, these formulas work for any quadrilateral. Unlike other general formulas of a quadrilateral, there is no need to calculate angles or other distances in the quadrilateral, it is sufficient to have only the lengths of the sides and the lengths of the diagonals. We also provide *Bretschneider's formula* to find the area of a quadrilateral, this formula works for any quadrilateral using only the side lengths and the sum of two opposite angles.

Area of a circle, circumference, area of a sector. Area of a circle is given by the following formula $S = \pi \cdot r^2$, circumference of a circle $C = 2\pi \cdot r$, where r is the radius of the circle.

Area of a regular hexagon.
$$S = \frac{3\sqrt{3}a^2}{2},$$
where a is the side length.

Area of a regular polygon.
$$S = \frac{na^2}{4\tan\frac{180°}{n}} = \frac{hp}{2},$$
where n is the number of sides, a is the side length, h is the apothem, p is the perimeter.

1.2.2 Tangential, cyclic, bicentric and extangential quadrilaterals, Pitot's theorem.

Tangential quadrilateral (or circumscribed quadrilateral). A tangential quadrilateral (or circumscribed quadrilateral) is a convex quadrilateral whose all four sides are tangent to a single circle within the quadrilateral.

Remark (incircle and inradius). This circle is called *incircle* and its radius is called *inadius* of the quadrilateral.

Pitot's theorem. A quadrilateral is tangential if and only if the sums of lengths of opposite sides are the same.

The direct statement of this theorem was proved by a French scientist *Henri Pitot*, the converse statement was proved later on by a Swiss mathematician *Jakob Steiner*.

Tangential quadrilateral theorem. A convex quadrilateral is tangential if and only if the angle bisectors of its four angles intersect at one point (are concurrent). Moreover, the intersection point is the center of the incircle (incenter).

Tangent lengths. Eight *tangent lengths* of a tangential quadrilateral are the line segments from a vertex to the points of tangency of its incircle and its sides.

Tangent lenghts theorem. In a tangential quadrilateral two tangent lengths corresponding to each vertex are congruent to each other (are equal in lenght).

Remark. This theorem is equivalent to the following statement: two tangents to a circle from a given point are equal in lenght to where they touch the circle (points of tangency).

The area S of a tangential quadrilateral can be given by different formulas, the simplest one is the following formula:
$$S = r \cdot s,$$
where s is the semiperimeter of the quadrilateral and r is the inradius.

Remark. This formula holds true for any tangential polygon.

Note that, if we denote the opposite sides of the tangential quadrilateral by a, c and b, d, then from the definition of a semi-perimeter and from Pitot's theorem we have that
$$s = \frac{a+b+c+d}{2} = a+c = b+d.$$

Remark (alternative formula of the area of a tangential quadrilateral.) Above mentioned formula is not the only formula of the area of a tangential quadrilateral. Its area can be given by different formulas, in the subsection *Bretschneider's formula, diagonals and sides area formula of a quadrilateral* we provide another useful formula of the area of a tangential quadrilateral, where the area can be calculated using only the lenghts of its diagonals and sides.

Cyclic quadrilateral (or inscribed quadrilateral). A cyclic quadrilateral (or inscribed quadrilateral) is a quadrilateral whose all vertices lie on one circle.

Remark (circumcircle and circumradius). This circle is called *circumcircle* and its radius is called *circumradius* of the quadrilateral.

Cyclic quadrilateral theorem 1. A convex quadrilateral is cyclic if and only if its opposite angles are supplementary (they add up to 180°).

The direct statements of theorem 1, theorem 2 and theorem 3 were given by a prominent mathematician of Greco-Roman antiquity *Euclid* (Eukleides of Alexandria, reffered as one of the founders of Geometry).

Consequence. A convex quadrilateral is cyclic if and only if an exterior angle is equal to the opposite interior angle.

Cyclic quadrilateral theorem 2. A convex quadrilateral is cyclic if and only if the angle between a side and a diagonal is equal to the angle between the opposite side and the other diagonal.

As we have already mentioned the direct statement of this theorem was given by *Euclid*, the converse statement was proved later on by a French mathematician *Jacques Hadamard*.

Cyclic quadrilateral theorem 3. Let $ABCD$ be a convex quadrilateral, such that lines AB and CD intersect at point E, then $ABCD$ is cyclic if and only if

$$AE \cdot EB = DE \cdot EC.$$

Cyclic quadrilateral theorem 4. A convex quadrilateral is cyclic if and only if four perpendicular bisectors to its sides intersect at the same point (are concurrent). Note that this intersection point is the circumcenter of the quadrilateral.

The area of a cyclic quadrilateral can be given by different formulas, the simplest one is Brahmagupha's formula (see the next subsection).

Bicentric quadrilateral. A bicentric quadrilateral is a convex quadrilateral that is both tangential and cyclic (has simultaneously an incircle and a circumcircle).

Extangential quadrilateral. Extangential quadrilateral is a convex quadrilateral whose extensions of all four sides are tangent to a circle outside the quadrilateral.

Remark (excircle and exradius.) This circle is called *excircle* and its radius is called *exradius* of the quadrilateral.

Extangential quadrilateral theorem. A convex quadrilateral is extangential if and only if the sum of its some two adjacent sides is equal to the sum of the other two sides.

This theorem was proved by a Swiss mathematician *Jakob Steiner*. Note that there are some other characterization theorems for extangential quadrilaterals (related to its angles), but we do not cover them in this book, as it is outside the scope of AMC.

1.2.3 Heron's formula and Brahmagupha's formula.

Heron's formula. *Heron's formula* is named after a well-known mathematician and experimenter of Greco-Roman antiquity *Hero of Alexandria*, also called *Heron of Alexandria*. This formula is applied to find the area S of any triangle

$$S = \sqrt{s(s-a)(s-b)(s-c)},$$

where a, b, c are the side lengths of given triangle and s is the semi-perimeter of the triangle, that is

$$s = \frac{a+b+c}{2}.$$

Brahmagupta's formula. *Brahmagupta's formula* is named after an Indian mathematician and astronomer Brahmagupta. This formula is applied to find the area S of any *cyclic* quadrilateral

$$S = \sqrt{(s-a)(s-b)(s-c)(s-d)},$$

where a, b, c, d are the side lengths of given cyclic quadrilateral and s is the semi-perimeter of the quadrilateral, that is

$$s = \frac{a+b+c+d}{2}.$$

1.2.4 Ceva's theorem, Menelaus' theorem, Stewart's theorem, Ptolemy's theorem.

Ceva's theorem. Given a triangle ABC. Let points D, E, F be on sides BC, AC, AB, respectively. Line segments AD, BE, CF intersect at one point (are concurrent), if and only if

$$\frac{AF}{FB} \cdot \frac{BD}{DC} \cdot \frac{CE}{EA} = 1.$$

Ceva's theorem is named after an Italian mathematician *Giovanni Ceva*. Nevertheless, it was proved much earlier by other authors too.

Remark (cevians). Line segments AD, BE, CF are known as *cevians*.

Menelaus' theorem. Given a triangle ABC. Let D, E be given points on sides BC, AC, respectively, and F be a given point on line AB (outside side AB). Points D, E, F lie on the same line (are collinear), if and only if

$$\frac{FA}{FB} \cdot \frac{DB}{DC} \cdot \frac{EC}{EA} = 1.$$

Menelaus' theorem is named after a mathematician of Greco-Roman antiquity *Menelaus of Alexandria*.

Stewart's theorem. Given a triangle ABC. Let D be a given point on side BC, then

$$AD^2 = \frac{AB^2 \cdot CD + AC^2 \cdot BD}{BC} - BD \cdot CD.$$

Stewart's theorem is named after a Scottish mathematician *Mattew Stewart* and it is used to express the length of a cevian in a triangle by the lengths of the sides.

Ptolemy's theorem. Quadrilateral $ABCD$ is a cyclic quadrilateral if and only if

$$AB \cdot CD + BC \cdot AD = AC \cdot BD.$$

Ptolemy's theorem is named after a mathematician of Greco-Roman antiquity *Claudius Ptolemaeus*, also known as *Ptolemy (of Alexandria)*.

Remark (Ptolemy's inequality). Let $ABCD$ be a quadrilateral, then

$$AB \cdot CD + BC \cdot AD \geq AC \cdot BD,$$

where the equality holds true if and only if $ABCD$ is a cyclic quadrilateral.

1.2.5 Bretschneider's formula, *diagonals and sides* area formula of a quadrilateral.

In this subsection we provide area formulas for a quadrilateral. The first formula we call *diagonals and sides* area formula of a quadrilateral, as for finding the area we use the lengths of its diagonals and sides.

***Diagonals and sides* area formula of a quadrilateral.** The area of any convex quadrilateral is given by the following general formula

$$S = \frac{1}{4}\sqrt{4e^2 f^2 - (b^2 + d^2 - a^2 - c^2)^2},$$

where S is the area, e, f are the lenghts of the diagonals and a, b, c, d are the lengths of the sides.

Alternative formulation of the (*diagonals and sides*) area formula of a quadrilateral.

$$S = \sqrt{(s-a)(s-b)(s-c)(s-d) - \frac{1}{4}(ac + bd + ef)(ac + bd - ef)},$$

where S is the area, s is the semi-perimeter and a, b, c, d are the lengths of the sides of the quadrilateral.

Another alternative but equivalent formula written in trigonometric form is the following one.
Bretschneider's formula.

$$S = \sqrt{(s-a)(s-b)(s-c)(s-d) - abcd\cos^2\left(\frac{\alpha+\gamma}{2}\right)},$$

where S is the area, s is the semi-perimeter, a, b, c, d are the lengths of the sides and α, γ are two opposite angles of the quadrilateral.

This result was published by a German mathematician *C. A. Bretschneider*. Besides Bretschneider this result was published by other authors too. The first two results were published by several authors also, in particular by a British mathematician *E.W. Hobson* and the alternative but equivalent formulation by the American mathematician *J.L. Coolidge*.

Cyclic quadrilateral area formula. Note that for a cyclic quadrilateral using either Bretschneider's formula with the property $\alpha + \gamma = 180°$ or the alternative formulation of *diagonals and sides* area formula with Ptolemy's theorem $ac + bd = ef$ one can easily deduce Brahmagupta's formula.

$$S = \sqrt{(s-a)(s-b)(s-c)(s-d)}.$$

Using the *diagonals and sides* area formula of a quadrilateral and Pitot's theorem $(a + c = b + d)$, we have deduced the following formula of the area for any tangential quadrilateral.

Tangential quadrilateral area formula. The area of a tangential quadrilateral can given by the following formula.

$$S = \frac{1}{2}\sqrt{e^2 f^2 - (ac - bd)^2}.$$

1.2.6 Parameshvara's formula for circumradius.

Parameshvara's formula The circumradius R of a cyclic quadrilateral can be given by the following formula

$$R = \frac{1}{4}\sqrt{\frac{(ab+cd)(ac+bd)(ad+bc)}{(p-a)(p-b)(p-c)(p-d)}}$$

where a, b, c, d are the lengths of the sides and p is the semi-perimeter of given cyclic quadrilateral.
Note that, using Brahmagupta's formula we can rewrite Parameshvara's formula in the following form

$$R = \frac{\sqrt{(ab+cd)(ac+bd)(ad+bc)}}{4 \cdot S},$$

where S is the area of given cyclic quadrilateral.
Parameshvara's formula is named after an Indian mathematician *Vatasseri Parameshvara Nambudiri*.

1.2.7 Formulas for volume and surface areas of three-dimensional shapes.

Volume and surface area of a cube.
$$V = a^3,$$
$$S = 6a^2,$$
where a is the side length of the cube.

Volume and surface area of a cone.
$$V = \frac{\pi r^2 h}{3},$$
$$S = \pi r^2 + \pi r l,$$
where r is the radius of the circular base, h is the height of the cone, l is the slant height.

Volume and surface area of a tetrahedron/pyramid.
$$V = \frac{Ah}{3},$$
where A is the area of the base, h is the height of the tetrahedron/pyramid.

Regular pyramid. A regular pyramid is a pyramid whose base is a regular polygon and whose lateral edges are all equal in length.
In the case of a regular pyramid its total surface area can be given by the following formula.
$$S = \frac{a \cdot l}{2} \cdot n + A = \frac{a \cdot l}{2} \cdot n + \frac{n \cdot a^2}{4 \tan \frac{180°}{n}},$$
where a is the lenght of the side, l is the slant height, n is the number of sides of the base (base is a regular polygon), A is the area of the base.

Volume and surface area of a cylinder.
$$V = \pi r^2 h,$$
$$S = 2\pi r^2 + 2\pi r h,$$
where r is the radius of the circular base, h is the height.

Volume and surface area of a rectangular prism.
$$V = lwh,$$
$$S = 2(lw + lh + wh),$$
where l is the length, w is the width, h is the height.

1.2.8 Trigonometric identities.

Some problems in the AMC 10, AMC 12 and many other math competitions may be solved using trigonometry. Mostly either by application of basic trigonometric identities or using the values of sines and cosines of common angles, such as

$$\sin 30° = \sin \frac{\pi}{6} = \cos 60° = \frac{1}{2},$$

$$\sin 45° = \sin \frac{\pi}{4} = \cos 45° = \frac{\sqrt{2}}{2}.$$

For more values of sines and cosines of common angles see *trigonometric unit circle* provided below.

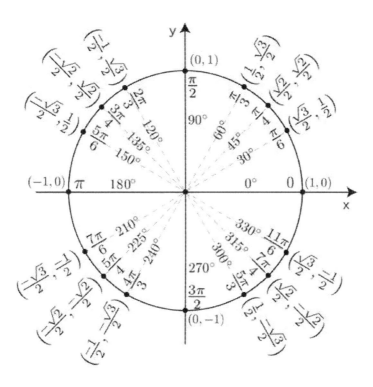

Quotient identities.

$$\tan \alpha = \frac{\sin \alpha}{\cos \alpha}.$$

$$\cot \alpha = \frac{\cos \alpha}{\sin \alpha}.$$

Reciprocal identities.

$$\cot \alpha = \frac{1}{\tan \alpha}.$$

$$\csc \alpha = \frac{1}{\sin \alpha}.$$

$$\sec \alpha = \frac{1}{\cos \alpha}.$$

Pythagorean identities.

$$\sin^2 \alpha + \cos^2 \alpha = 1.$$

$$\tan^2 \alpha + 1 = \frac{1}{\cos^2 \alpha}.$$

$$\cot^2 \alpha + 1 = \frac{1}{\sin^2 \alpha}.$$

Sum-to-product identities.

$$\sin \alpha + \sin \beta = 2 \sin \frac{\alpha + \beta}{2} \cos \frac{\alpha - \beta}{2}.$$

$$\sin \alpha - \sin \beta = 2 \sin \frac{\alpha - \beta}{2} \cos \frac{\alpha + \beta}{2}.$$

$$\cos \alpha + \cos \beta = 2 \cos \frac{\alpha + \beta}{2} \cos \frac{\alpha - \beta}{2}.$$

$$\cos \alpha - \cos \beta = -2 \sin \frac{\alpha + \beta}{2} \sin \frac{\alpha - \beta}{2}.$$

Product-to-sum identities.

$$\sin \alpha \sin \beta = \frac{1}{2}(\cos(\alpha - \beta) - \cos(\alpha + \beta)).$$

$$\cos \alpha \cos \beta = \frac{1}{2}(\cos(\alpha - \beta) + \cos(\alpha + \beta)).$$

$$\sin \alpha \cos \beta = \frac{1}{2}(\sin(\alpha - \beta) + \sin(\alpha + \beta)).$$

Sum identities.

$$\sin(\alpha + \beta) = \sin \alpha \cos \beta + \cos \alpha \sin \beta.$$

$$\cos(\alpha + \beta) = \cos \alpha \cos \beta - \sin \alpha \sin \beta.$$

$$\tan(\alpha + \beta) = \frac{\tan \alpha + \tan \beta}{1 - \tan \alpha \tan \beta}.$$

Difference identities.

$$\sin(\alpha - \beta) = \sin \alpha \cos \beta - \cos \alpha \sin \beta.$$

$$\cos(\alpha - \beta) = \cos \alpha \cos \beta + \sin \alpha \sin \beta.$$

$$\tan(\alpha - \beta) = \frac{\tan \alpha - \tan \beta}{1 + \tan \alpha \tan \beta}.$$

Cofunction identities.

$$\sin\left(\frac{\pi}{2} - \alpha\right) = \cos \alpha.$$

$$\cos\left(\frac{\pi}{2} - \alpha\right) = \sin \alpha.$$

$$\tan\left(\frac{\pi}{2} - \alpha\right) = \cot \alpha.$$

$$\cot\left(\frac{\pi}{2} - \alpha\right) = \tan \alpha.$$

$$\csc\left(\frac{\pi}{2} - \alpha\right) = \sec \alpha.$$

$$\sec\left(\frac{\pi}{2} - \alpha\right) = \csc \alpha.$$

Negative angle (even/odd) identities.

$$\sin(-\alpha) = -\sin \alpha.$$

$$\cos(-\alpha) = \cos \alpha.$$

$$\tan(-\alpha) = -\tan \alpha.$$

$$\cot(-\alpha) = -\cot \alpha.$$

$$\csc(-\alpha) = -\csc \alpha.$$

$$\sec(-\alpha) = \sec \alpha.$$

Double-angle identities.

$$\sin 2\alpha = 2 \sin \alpha \cos \alpha.$$

$$\cos 2\alpha = \cos^2 \alpha - \sin^2 \alpha = 2\cos^2 \alpha - 1 = 1 - 2\sin^2 \alpha.$$

$$\tan 2\alpha = \frac{2 \tan \alpha}{1 - \tan^2 \alpha}.$$

Half-angle identities.

$$\sin\frac{\alpha}{2} = \pm\sqrt{\frac{1-\cos\alpha}{2}}.$$

$$\cos\frac{\alpha}{2} = \pm\sqrt{\frac{1+\cos\alpha}{2}}.$$

$$\tan\frac{\alpha}{2} = \frac{\sin\alpha}{1+\cos\alpha} = \frac{1-\cos\alpha}{\sin\alpha}.$$

Triple angle identities.

$$\sin 3\alpha = 3\sin\alpha - 4\sin^3\alpha.$$

$$\cos 3\alpha = 4\cos^3\alpha - 3\cos\alpha.$$

$$\tan 3\alpha = \frac{3\tan\alpha - \tan^3\alpha}{1 - 3\tan^2\alpha}.$$

Symmetry identities.

$$\sin(\pi - \alpha) = \sin\alpha.$$
$$\cos(\pi - \alpha) = -\cos\alpha.$$
$$\tan(\pi - \alpha) = -\tan\alpha.$$
$$\cot(\pi - \alpha) = -\cot\alpha.$$
$$\csc(\pi - \alpha) = \csc\alpha.$$
$$\sec(\pi - \alpha) = -\sec\alpha.$$
$$\sin(\pi + \alpha) = -\sin\alpha.$$
$$\cos(\pi + \alpha) = -\cos\alpha.$$
$$\tan(\pi + \alpha) = \tan\alpha.$$
$$\cot(\pi + \alpha) = \cot\alpha.$$
$$\csc(\pi + \alpha) = -\csc\alpha.$$
$$\sec(\pi + \alpha) = -\sec\alpha.$$

1.2.9 Complex numbers, de Moivre's formula, Euler's formula.

Complex numbers. A complex number is a number that can be expressed in the form $a + bi$, where a, b are real numbers and i is an indeterminate satisfying $i^2 = -1$. As no real number satisfies this equation, i is called an *imaginary number*, a is called the *real part* and b is called the *imaginary part*.

Remark. The real part of a complex number z is denoted by $Re(z)$, the imaginary part of a complex number z is denoted by $Im(z)$.

Example. Consider complex number $5 + 3i$, then $Re(5 + 3i) = 5$ and $Im(5 + 3i) = 3$.

Absolute value (modulus or magnitude) of a complex number. The *absolute value* (modulus or magnitude) of a complex number $z = a + bi$ is denoted by $|z|$ and is equal to:

$$|z| = \sqrt{a^2 + b^2}.$$

Conjugate of a complex number. The complex *conjugate* of complex number $z = a + bi$ is given by $a - bi$ and is denoted by $\bar{z}$.

Remark. Note that
$$z \cdot \bar{z} = (a + bi)(a - bi) = a^2 + b^2 = |z|^2 = |\bar{z}|^2.$$

De Moivre's formula. For any real number x and integer n it holds that

$$(\cos x + i \sin x)^n = \cos(nx) + i\sin(nx),$$

where i is the imaginary number.

This formula is named after a French mathematician *Abraham de Moivre*. De Moivre's formula can be considered as a consequence of the below mentioned *Euler's formula*.

Euler's formula. Euler's formula establishes the following relationship between the trigonometric functions and the complex exponential function:

$$e^{ix} = \cos x + i\sin x,$$

where e is called *Euler's number* after a Swiss mathematician *Leonhard Euler* and i the imaginary number. Mathematical constant e is approximately equal to 2.71828 and is the base of the natural logarithm.

Remark. The expression $\cos x + i \sin x$ sometimes is denoted as $cis(x)$.

1.3 Number theory (AMC 12): the most useful formulas and theorems

1.3.1 Unique-prime-factorization theorem (fundamental theorem of arithmetic).

Unique-prime-factorization theorem (fundamental theorem of arithmetic). Every integer greater than 1 is either is prime number or can be represented as the product of prime numbers. Moreover, this representation is unique, up to the order of the factors.

Mathematically, this statement can be rewritten in the following way.

Unique-prime-factorization theorem (fundamental theorem of arithmetic). Every integer n greater than 1 can be represented as

$$n = p_1^{\alpha_1} \cdot p_2^{\alpha_2} \cdot \ldots \cdot p_k^{\alpha_k},$$

where $p_1, p_2, \ldots, p_k$ are distinct primes and $\alpha_1, \alpha_2, \ldots, \alpha_k$ are positive integers.

Remark. This representation is called the **canonical representation** of n or **standard form** of n.

1.3.2 Number of divisors of a composite number, sum and product of divisors.

Number of divisors of a composite number n. If the prime-factorization of a composite number n is $n = p_1^{\alpha_1} \cdot p_2^{\alpha_2} \cdot \ldots \cdot p_k^{\alpha_k}$, then the number of divisors of n (denoted by $d(n)$) is equal to

$$d(n) = (\alpha_1 + 1)(\alpha_2 + 1)\ldots(\alpha_k + 1).$$

Sum of divisors of a composite number. If the prime-factorization of a composite number n is $n = p_1^{\alpha_1} \cdot p_2^{\alpha_2} \cdot \ldots \cdot p_k^{\alpha_k}$, then the sum of divisors of n (denoted by $\tau(n)$) is equal to

$$\tau(n) = \frac{p_1^{\alpha_1+1} - 1}{p_1 - 1} \cdot \frac{p_2^{\alpha_2+1} - 1}{p_2 - 1} \cdot \ldots \cdot \frac{p_k^{\alpha_k+1} - 1}{p_k - 1}.$$

Product of divisors of a composite number. If the prime-factorization of a composite number n is $n = p_1^{\alpha_1} \cdot p_2^{\alpha_2} \cdot \ldots \cdot p_k^{\alpha_k}$, then the product of divisors of n (denoted by $\pi(n)$) is equal to

$$\pi(n) = n^{\frac{d(n)}{2}},$$

where as mentioned above $d(n)$ is the number of divisors of n.

1.3.3 One useful lemma.

The following lemma can be applied as a very useful technique to treat non-standard *number theory* problems.

One useful lemma. Let n be a positive integer. Let $d_1, d_2, \ldots, d_k$ be all positive integer divisors of n, such that $1 = d_1 < d_2 < \ldots < d_k = n$, then

$$d_1 = \frac{n}{d_k}, d_2 = \frac{n}{d_{k-1}}, \ldots, d_k = \frac{n}{d_1}.$$

1.3.4 Modular arithmetic and congruence relation.

Congruence relation. For a positive integer n integers a and b are called *congruent modulo n* if $n \mid a-b$.
Remark. This notation means that n divides $a - b$. In other words it means that $a - b$ is divisible by n, that is, if there exits an integer k such that $a - b = k \cdot n$.
Congruence modulo n is denoted in the following way.

$$a \equiv b \ (mod \ n).$$

1.3.5 Fermat's little theorem and Wilson's theorem.

Fermat's little theorem. If p is prime number, then for any integer a, we have that

$$p \mid a^p - a.$$

Remark. Using *modular arithmetic* and *congruence relation* Fermat's little theorem can be rewritten in the following way

$$a^p \equiv a \ (mod \ p).$$

Fermat's little theorem (alternative formulation). If p is prime number, then for any integer a, we have that

$$p \mid a^{p-1} - 1.$$

Remark. Using *modular arithmetic* and *congruence relation* alternative formulation of Fermat's little theorem can be rewritten in the following way

$$a^{p-1} \equiv 1 \ (mod \ p).$$

Fermat's little theorem is named after a prominent French mathematician and lawyer *Pierre de Fermat*.
Wilson's theorem. Integer number n greater than 1, is prime number if and only if

$$n \mid (n-1)! + 1.$$

Remark. Using *modular arithmetic* and *congruence relation* Wilson's theorem can be rewritten in the following way

$$(n-1)! \equiv -1 \ (mod \ n).$$

Wilson's theorem is named after an English mathematician *John Wilson*. It is funny, but Wilson was neither the first person to state this theorem nor the first person to prove it. A prominent French mathematician of Italian descent *Joseph-Louis Lagrange* (born *Giuseppe Luigi Lagrangia*) gave the first published proof of this theorem. Several other mathematicians have stated this result earlier, without providing a published proof.

1.4 Combinatorics and probability (AMC 12): the most useful formulas and theorems

1.4.1 Rule of sum and rule of product.

In combinatorics, the rule of sum (or addition principle) and the rule of product (or multiplication principle) are basic counting principles.

The rule of sum (or addition principle). Let n and m be nonnegative integers. If there are n choices for one action and m choices for another action and these two actions cannot be done simultaneously, then there are $n + m$ ways to perform one of these actions.

The rule of product (or multiplication principle). Let n and m be nonnegative integers. If there are n choices for one action and m choices for another action, then there are $n \cdot m$ ways to perform both of these nonnegative actions.

1.4.2 Permutations.

Permutation. A permutation of a set of objects is an ordering (rearrangement) of these objects.

Permutation with repetition (or r-tuple). Let r and n be nonnegative integers, such that $n \geq r$. A permutation with repetition (or r-tuple) is an ordered selection of r elements from a set of n elements, where repetition is allowed.

Theorem (the number of permutations with repetition). Let r and n be nonnegative integers, such that $n \geq r$. The number of permutations with repetition for selection r elements from a set of n elements where repetition is allowed is equal to n^r.

Permutation without repetition. Let k and n be nonnegative integers, such that $n \geq k$. A permutation without repetition is an ordered selection of k elements from a set of n elements, where repetition is not allowed.

Theorem (the number of permutations without repetition). Let k and n be nonnegative integers, such that $n \geq k$. The number of permutations without repetition of obtaining an ordered subset of k elements from a set of n elements is denoted by nP_k.

$$nP_k = \frac{n!}{(n-k)!} = n(n-1) \cdot \ldots \cdot (n-(k-1)),$$

where $n! = 1 \cdot 2 \cdot \ldots \cdot n$.

Corollary. Let n be a positive integer. The number of permutations without repetition of obtaining an ordered subset of n elements from a set of n elements is denoted by P_n and

$$P_n = n!.$$

1.4.3 Combinations.

Combination. Let k and n be nonnegative integers, such that $n \geq k$. The number of ways of obtaining an unordered subset of k elements from a set of n elements is called combination and is denoted by $\binom{n}{k}$ or C_n^k or nC_k.

Remark 1 (binomial coeffecint). The symbol $\binom{n}{k}$ is called the binomial coefficient.

Remark 2. The symbols $\binom{n}{k}$ or nC_k or C_n^k are read as n *choose* k, meaning that there are $\binom{n}{k}$ ways to choose an unordered subset of k elements from a fixed set of n elements.

Remark 3. One can say that a permutation is an "ordered combination."

Theorem (the number of combinations.) Let k and n be nonnegative integers, such that $n \geq k$. The number of combinations of obtaining an unordered subset of k elements from a set of n elements can be found by the following formula.

$$\binom{n}{k} = \frac{n!}{k!(n-k)!}.$$

1.4.4 Stars and bars technique (integer equations).

In combinatorics very often arise problems such as counting the number of ways to group identical objects, for example placing indistinguishable balls into ennumerated urns, or finding the number of nonnegative integer solutions an equation has. Stars and bars technique is a short and elegant way to treat such problems. We provide the following example to explain how this technique works.

Example. Consider the following equation

$$a + b + c + d + e = 16,$$

where a, b, c, d, e are nonnegative integers. What is the total number of all nonnegative integer solutions of this equation?

Solution. Assume there are 20 places, where we place 16 stars and 4 bars (one per place). The main idea is that any such arrangment represents a solution of given equation. For example

$$** \mid * * ** \mid * * * \mid * * * * ** \mid *$$

represents the solution

$$2 + 4 + 3 + 6 + 1 = 16.$$

Remark. Note that before the first bar and after the last bar it is possible to place 0 stars. Therefore, given problem is equivalent to the following question: *In how many ways is it possible to put 16 stars and 4 bars in 20 places?*. It is equivalent to fixing 4 places out of 20 places and putting stars in the empty places (one star per place). This can be done by $\binom{20}{4}$ ways. Using the corresponding formula for the binomail coefficient, we obtain that

$$\binom{20}{4} = 4845.$$

1.4.5 Probability.

Probability is a numerical description of how likely an event is to occur (or how likely is that a proposition holds true).

Probability is a number from the interval $[0, 1]$, where 0 stands for impossibility and 1 stands for certainty.

Probability range. $0 \leq P(event) \leq 1$.

Theoretical probability expresses the likelihood that something will occur.

Theoretical probability is equal to the number of favorable outcomes divided by the total number of possible outcomes.

Theoretical probability formula.

$$P(event) = \frac{number\ of\ favorable\ outcomes}{number\ of\ possible\ outcomes}.$$

Complementary events. Complementary probability is the probability of a given event not occurring (or of a different event occurring that can only occur if the first event does not occur).

Complementary events formula. Let A be a given event, we denote by A^C its complementary event. Then, we have that

$$P(A^C) + P(A) = 1.$$

For example, the probability of a tossed coin landing on heads is $\frac{1}{2}$ and the complementary probability of the tossed coin not landing on heads (landing on tails) is

$$1 - \frac{1}{2} = \frac{1}{2}.$$

Independent events. Two events are called independent if the fact that one event occurs does not affect the probability of the other event occurring.

For example, if someone tosses a coin, then the possible result (head or tail) is not affected by previous tosses.

Independent events formula. Events A and B are independent if and only if

$$P(A \text{ and } B) = P(A \cap B) = P(A) \cdot P(B).$$

For example, if two coins are tossed the probability of both being heads is

$$\frac{1}{2} \cdot \frac{1}{2} = \frac{1}{4}.$$

Disjoint events. Two events are called disjoint if they never occur simultaneously. In the literature, they are also called mutually exclusive events.

Disjoint events formula. If A and B are disjoint events, then the probability that both of them occur simultaneously is:

$$P(A \cap B) = 0.$$

Rule of addition of probability. If A and B are two events in a probability experiment, then the probability that either one of them occurs is:

$$P(A \cup B) = P(A) + P(B) - P(A \cap B).$$

Obviously, if A and B are disjoint events, then we have that:

Rule of addition of probability for disjoint events.

$$P(A \cup B) = P(A) + P(B).$$

Chapter 2

AMC 12 type practice tests

2.1 AMC 12 type practice test 1

Problem 2.1. *Alice needs to solve exactly 100 problems in n days. Given that each day she can solve either 5, 6 or 7 problems. What is the smallest possible value of n?*

(A) 14   (B) 15   (C) 20   (D) 16   (E) 17

Problem 2.2. *The sum of the squares of two positive numbers is three times their product. How many times is the square of the sum of these numbers greater than the square of their difference?*

(A) 5   (B) 3   (C) 9   (D) 2   (E) 4

Problem 2.3. *A student was assigned a test consisting of 10 problems. Each correct answer was scored 2 points, an incomplete answer was scored 1 point, and no point was scored for a wrong answer. In the end of the test the total score of the student was 19. Which of the following statements is true?*

(A) The student correctly answered to all 10 problems.
(B) At least one of the answers was incorrect.
(C) All the answers for all 10 problems were incomplete.
(D) Only one answer was correct.
(E) Only one answer was incomplete.

Problem 2.4. *By what percent is the circumference of the incircle of a square less than the perimeter of that square?*

(A) 50   (B) 22   (C) 21   (D) $\dfrac{100(4-\pi)}{\pi}$ (E) $25(4-\pi)$

Problem 2.5. *Participants of a mathematics conference stay in two hotels. Participants staying in the same hotel shook hands with each other exactly once, while participants staying in different hotels did not. The total number of handshakes is the product of the number of participants in each hotel. Given that the total number of participants is greater than 17 and less than 34. What is the total number of participants?*

(A) 20   (B) 25   (C) 18   (D) 30   (E) 33

Problem 2.6. *Let $ABCD$ be a square of side length 12. Let M be an interior point of $ABCD$. Given that the distance from M to side AB, AD, CD is a, b, c, respectively. How many possible points M are there, such that a, b, c are integers and there exists a triangle with sides a, b, c?*

(A) 72   (B) 60   (C) 66   (D) 59   (E) 61

Problem 2.7. *Let n be a positive integer and function $f(n)$ be defined as follows:*

$$f(1) = 5,$$

$$f(n) = f\left(\frac{n-1}{2}\right) + 1, if \ n \ is \ odd \ greater \ than \ 1,$$

$$f(n) = f\left(\frac{n}{2}\right), if \ n \ is \ even.$$

What is the value of $f(2017)$?

(A) 9   (B) 11   (C) 10   (D) 1008   (E) 20

Problem 2.8. *Given a cube with an edge length of 1 unit. Let Φ be the solid formed by all points that are located at a distance not greater than 1 unit from some point on the surface of the cube. What is the volume of Φ?*

(A) 27   (B) $7 + 3\pi$ (C) $7 + \dfrac{4\pi}{3}$ (D) $7 + \dfrac{13\pi}{3}$ (E) $6 + 4\pi$

Problem 2.9. *What is the locus of all possible points (x, y) satisfying the following inequality?*

$$|x + y - 1| + |x - y + 1| + |x + y + 1| + |x - y - 1| \le 4$$

(A) four vertices of a square.
(B) a square and its inner region.
(C) a triangle and its inner region.
(D) three vertices of a triangle.
(E) eight points.

Problem 2.10. *Let the distance between the centers of two circles of radii 8 and 2 be equal to 3. A circle of radius 1 is randomly placed within the circle of radius 8. What is the probability that the circles of radii 1 and 2 intersect?*

(A) $\dfrac{1}{16}$ (B) $\dfrac{9}{64}$ (C) $\dfrac{9}{49}$ (D) $\dfrac{1}{5}$ (E) $\dfrac{1}{2}$

Problem 2.11. *Let some internal angles of a convex polygon be acute. Given that the sum of its non-acute angles is equal to $2017°$ and the sum of its acute angles is equal to $n°$. What is the value of n?*

(A) 323   (B) 117   (C) 143   (D) 163   (E) 37

Problem 2.12. *A positive number is called "nice", if at least six of its divisors are from the set*

$$\{1, 2, 3, 4, 5, 6, 7, 8, 9, 10\}.$$

What is the smallest possible value of the positive difference of two "nice" numbers?

(A) 1   (B) 3   (C) 6   (D) 4   (E) 2

Problem 2.13. *Sophia planned to cover a distance of 160 miles at a constant rate. Given that half of the distance she moved 5 miles per hour faster than she has planned, and for the rest of the distance she moved 4 miles per hour slower than she has planned. It turned out that Sophia has spent as much time covering the total distance as she has planned. In how much time (in hours) Sophia has planned to cover the total distance?*

(A) 4   (B) 8   (C) 3   (D) 2   (E) 5

Problem 2.14. *Let five chairs be placed around a circular table. In how many different ways can two girls and three boys be seated, such that two girls do not sit next to each other?*

(A) 120 (B) 60 (C) 30 (D) 20 (E) 24

Problem 2.15. *What is the smallest integer value of n greater than 1, such that the following inequality holds true?*
$$\sin n + 2\cos n + \tan n > 0.$$

(A) 2 (B) 3 (C) 4 (D) 5 (E) 6

Problem 2.16. *Let circles of radii 1, 2, 3 be pairwise externally tangent. What is the radius of the circle that is internally tangent to each of the three circles?*

(A) $2\frac{5}{23}$ (B) $4\frac{10}{23}$ (C) $8\frac{20}{23}$ (D) 5 (E) 6

Problem 2.17. *What is the total number of all complex numbers z satisfying the following two conditions?*
$$z^{12} = 1,$$
$$(z - \sqrt{3})^{36} = 1.$$

(A) 0 (B) 2 (C) 6 (D) 4 (E) 3

Problem 2.18. *Let $n = \overline{a_1 a_2 ... a_k}$ and*
$$T(n) = |a_1 - a_2 + ... + (-1)^{k-1} a_k|.$$

For example
$$T(1237) = |1 - 2 + 3 - 7| = 5.$$

Given that $T(n) = 4$ for some positive integer n. Which of the following values can be equal to $T(n-1)$?

(A) 2 (B) 9 (C) 6 (D) 1 (E) 7

Problem 2.19. *Let a square with a side length of x be inscribed into a triangle with side lengths of 13, 14, 15, such that two of its vertices lie on the smallest side of the triangle. Let another square with a side length of y be inscribed into another triangle which is congruent to the first triangle, such that two of its vertices lie on the largest side of the triangle. What is the value of $\frac{15}{x} - \frac{13}{y}$?*

(A) $\frac{56}{195}$ (B) 1 (C) $\frac{1685}{1703}$ (D) $\frac{2}{3}$ (E) $\frac{3}{2}$

Problem 2.20. *For how many real numbers a the values of the following three expressions form a geometric sequence?*
$$\log_2 a, \log_2(a^2 - 1), -\log_2 \frac{(a^2-1)^{\sqrt{2}-1}}{a^{\sqrt{2}}}.$$

(A) 2 (B) 3 (C) 1 (D) 4 (E) 0

Problem 2.21. *Let a, b, c, d be integers, such that $ad \neq 0$ and 5 is a root of the polynomial $ax^3 + bx^2 + cx + d$. What is the smallest possible value of $|a| + |b| + |c| + |d|$?*

(A) 4 (B) 5 (C) 6 (D) 12 (E) 14

Problem 2.22. *A particle moves with a constant speed on a unit square grid. Every second, it moves from point (x, y) either to point $(x + 1, y)$ or to point $(x, y + 1)$, where x, y are nonnegative integers. The particle starts at point (0, 0) and ends its route at point (4,4). What is the probability that the route passes through point (2,2)?*

(A) $\dfrac{1}{2}$ (B) $\dfrac{3}{8}$ (C) $\dfrac{17}{35}$ (D) $\dfrac{18}{35}$ (E) $\dfrac{5}{8}$

Problem 2.23. *Given that there exists a rational number k, such that each of the polynomials*

$$x^3 + x^2 + kx + 2,$$

and

$$x^4 - 4x^3 + 4x^2 + (k+5)x - 3,$$

has an integer root larger than 1. What is the value of k?

(A) $\dfrac{5}{24}$ (B) $-\dfrac{3}{25}$ (C) -7   (D) 70   (E) 3

Problem 2.24. *Let $ABCD$ be a cyclic quadrilateral, such that $AB = 3, BC = 8, CD = 6$ and $AD = 4$. Let M and N be points on line BD, such that AM is parallel to CD and CN is parallel to AB. What is the value of $AM + CN$?*

(A) 24   (B) 25.5   (C) 18   (D) 27   (E) 13.5

Problem 2.25. *Consider the set*

$$V = \{-3 + 2i, 2 + i, -1\dfrac{1}{3} - 3i\}.$$

Let $S = z_1 + z_2 + ... + z_{10}$, where complex number z_j is an element of V for $j = 1, 2, ..., 10$. What is the probability that $S = 0$?

(A) $\dfrac{280}{6561}$ (B) $\dfrac{140}{6561}$ (C) $\dfrac{280}{2187}$ (D) $\dfrac{140}{2187}$ (E) $\dfrac{70}{729}$

2.2 AMC 12 type practice test 2

Problem 2.26. What is the value of the following expression?
$$(3^{-1} + 6^{-1} - 4^{-1})^{-1} : 5^0.$$

(A) $\dfrac{1}{4}$ (B) 4   (C) 5   (D) 1   (E) $\dfrac{4}{5}$

Problem 2.27. Let the ratio of the perimeter of a triangle to its largest side be a positive integer less than 4. Given that its largest side is equal to 24. What is the value of the perimeter of the triangle?

(A) 72   (B) 60   (C) 56   (D) 48   (E) 50

Problem 2.28. There are 7 boys and 8 girls in a class. The teacher gave each of these 15 students a test with 10 problems. Each problem is scored either 1 or 0 points. Given that the average score of all girls was 7 and the average score of the entire class was 5.6. What was the average score of all boys?

(A) 4.2   (B) 5   (C) 4   (D) 6   (E) 8

Problem 2.29. Let the ratio of the sum of the cubes of two positive numbers to the difference of their cubes be equal to $\dfrac{189}{61}$. By what percent is the larger number greater than the smaller number?

(A) 50   (B) 25   (C) 100   (D) 40   (E) 60

Problem 2.30. Given that $2 < a < 3, 3 < b < 4$ and $1 < c < 2$. Which of the following intervals contains the value of the expression $a - \dfrac{b}{c}$?

(A) (-4,-3)   (B) (1.5, 2)   (C) (2, 3)   (D) (3, 5)   (E) (-2, 1.5)

Problem 2.31. Five years ago Henry was three times Mia's age. Given that (now) the sum of their ages is equal to 34. How old is Mia?

(A) 6   (B) 10   (C) 13   (D) 11   (E) 8

Problem 2.32. Let points A, B lie in the first quadrant of the coordinate plane and belong to the graph of function $y = \dfrac{1}{x^3}$. The abscissa of point A is 25% greater than the abscissa of point B. By what percent is the ordinate of point A smaller than the ordinate of point B?

(A) 75   (B) 25   (C) 20   (D) 50   (E) 48.8

Problem 2.33. Let the ratio of the lengths of the legs of a right triangle be 5 : 12. Given that the diameter of its incircle is equal to D and the area of given triangle is equal to $m \cdot D^2$. What is the value of m?

(A) $\dfrac{15}{4}$ (B) $\dfrac{15}{8}$ (C) $\dfrac{25}{144}$ (D) $\dfrac{12}{5}$ (E) $\dfrac{5}{12}$

Problem 2.34. Carol randomly selects any two of the following numbers: 1, 2, 3, 4, 5, 6. Then, Claudia randomly selects any two numbers from the list of the remaining numbers. Finally, Cheryl selects the last two remaining numbers. What is the probability that one of the selected numbers of each girl is the multiple of the other selected number of the same girl?

(A) $\dfrac{1}{15}$ (B) $\dfrac{1}{4}$ (C) $\dfrac{1}{6}$ (D) $\dfrac{1}{10}$ (E) $\dfrac{1}{5}$

Problem 2.35. For how many positive integers n the value of $\dfrac{20n}{n+15}$ is an integer?

(A) 18 (B) 10 (C) 9 (D) 7 (E) 8

Problem 2.36. For any lines a, b, c in the plane let k be the number of all circles in that plane, such that each of them touches each of the lines a, b, c (note that the value of k depends on the choice of the lines a, b, c). How many possible values of k are there?

(A) 0 (B) 1 (C) 4 (D) 3 (E) 2

Problem 2.37. Two parabolas given by equations $y = x^2 + bx - 2$ and $y = -3x^2 + (b-4)x + 6$, intersect each other at exactly two points on the x-axis. What is the value of the area of a quadrilateral which has vertices at these intersection points and at the vertices of the parabolas.

(A) 27 (B) 13.5 (C) 13 (D) 12 (E) 10

Problem 2.38. More than one boy and more than one girl attend math lessons. During the first lesson, each girl and each boy exchanged with each other one of their photos. During the second lesson, every two students of the same gender exchanged with each other one of their photos. Given that the total number of exchanges during the first lesson was the same as the total number of exchanges during the second lesson. How many students attend math lessons?

(A) 3 (B) 5 (C) 7 (D) 4 (E) 10

Problem 2.39. What is the value of a for which the following equation holds true?

$$\log_2 3 \cdot \log_3 4 + \log_4 5 \cdot \log_5 6 \cdot \log_6 7 \cdot \log_7 a = 3.5.$$

(A) 4 (B) 5 (C) 6 (D) 7 (E) 8

Problem 2.40. What is the smallest positive value of n for which the number $\dfrac{n}{10!}$ can be expressed as a finite decimal?

(A) 405 (B) 189 (C) 567 (D) 135 (E) 199

Problem 2.41. Let $ABCD$ be a tetrahedron, such that $AB = 8, AC = 4, AD = 4, BC = \sqrt{34}, BD = 4\sqrt{3}$ and $CD = 5$. What is the volume of the tetrahedron?

(A) $\dfrac{128}{3}$ (B) $\sqrt{39}$ (C) $\dfrac{128}{6}$ (D) $8\sqrt{34}$ (E) $2\sqrt{39}$

Problem 2.42. Let a twenty-digit number starts with 2017 and its other digits are chosen randomly from the digits 0, 1, 2, 3, 4, 5, 6. What is the probability that such twenty-digit number is divisible by 140?

(A) $\dfrac{18}{343}$ (B) $\dfrac{15}{343}$ (C) $\dfrac{12}{343}$ (D) $\dfrac{1}{49}$ (E) $\dfrac{4}{343}$

Problem 2.43. For how many values of a in the interval $(-1, 1)$ does the quadratic expression $x^2 + ax + 3a + 2$ have at least one integer root?

(A) 2 (B) 5 (C) 101 (D) 0 (E) 1

Problem 2.44. Let $ABCD$ be a convex quadrilateral, such that $AB = 1, BC = 4, CD = 8, AD = 7$. What is the greatest possible value of the area of the quadrilateral?

(A) $2\sqrt{65}$ (B) 17 (C) 18 (D) 19 (E) 18.5

Problem 2.45. *Given a trapezoid with bases BC and AD, such that $AB = 4, BD = 3, AD = 5$ and $BC = 3$. What is the total number of all possible values of CD?*

(A) 1   (B) 3   (C) 4   (D) 2   (E) 5

Problem 2.46. *Let F_1 and F_2 be the foci of the ellipse given by equation $\dfrac{x^2}{25} + \dfrac{y^2}{9} = 1$. Let C be a point on the ellipse, such that $\angle F_1CF_2 = 90°$. What is the value of the area of triangle F_1CF_2?*

(A) 16   (B) 9   (C) 10   (D) 12   (E) 12.5

Problem 2.47. *We say that a sequence of 12 numbers is "nice" if its each term is either 1, 2 or 3, and if the the sum of any four consecutive terms is either 7 or 9. How many "nice" sequences are there?*

(A) 100   (B) 36   (C) 256   (D) 128   (E) 472

Problem 2.48. *Let ABC be an equilateral triangle with a side length of 2. Let points M and N be randomly chosen points on line segments AB and AC, respectively. What is the probability that $MN \leq \sqrt{3}$?*

(A) $\dfrac{\pi}{6}$ (B) $\dfrac{1}{2}$ (C) $\dfrac{1}{3}$ (D) $\dfrac{\pi\sqrt{3}}{6}$ (E) $\dfrac{6-\pi}{\pi}$

Problem 2.49. *Let r and s be randomly chosen rational numbers from the interval (0, 1). Each of them can be expressed in the form $\dfrac{n}{12}$, where n is an integer. What is the probability that the absolute value of the number $\cos(\pi r) + i\cos(\pi s)$ is equal to 1?*

(A) $\dfrac{19}{121}$ (B) $\dfrac{2}{11}$ (C) $\dfrac{20}{121}$ (D) $\dfrac{24}{121}$ (E) $\dfrac{28}{121}$

Problem 2.50. *Let parabola given by equation $y = x^2$ be drawn on a rectangular coordinate plane. Given a circle with the following properties: it passes through point (1, 1), it is tangent to the x-axis at a point that has a positive abscissa, the circle and the parabola have a common tangent line passing through point (1, 1). What is the radius of that circle?*

(A) $\dfrac{5-\sqrt{5}}{4}$ (B) $\dfrac{1}{2}$ (C) $\dfrac{2}{3}$ (D) $\dfrac{5-\sqrt{5}}{2}$ (E) $\dfrac{3-\sqrt{5}}{2}$

2.3 AMC 12 type practice test 3

Problem 2.51. Let $ABCD$ be a square with a side length of 6. Let E be a point on the line segment BC. What is the value of the area of triangle AED?

(A) 30   (B) 24   (C) 12   (D) 18   (E) 6

Problem 2.52. Among ten children on a hiking tour, any two of them have different number of candies. They split up equally into two groups, and it turns out that the total number of candies in the first group is five times smaller than the total amount of candies in the second group. What is the smallest possible total number of candies all children together can have?

(A) 70   (B) 50   (C) 60   (D) 40   (E) 30

Problem 2.53. There are 112 apples in one box, 97 apples in a second box, and 88 apples in a third box. First, a apples were transferred from the first box to the second box, and then b apples were transferred from the second box to the third box. After that, all the boxes had an equal number of apples. What is the value of $a + b$?

(A) 11   (B) 13   (C) 20   (D) 25   (E) 24

Problem 2.54. Given that $2^{2013} : 2^x = 2^{25} \cdot 2^{1975}$. What is the value of x?

(A) 0   (B) $\dfrac{2013}{2010}$ (C) 4013   (D) 13   (E) -13

Problem 2.55. Let in the classroom 22 students be seated in two rows. Given that 25% of the students in the first row and 20% of the students in the second row are enrolled in an after-school math program. How many students in the classroom are enrolled in the after-school math program?

(A) 4   (B) 2   (C) 5   (D) 3   (E) 6

Problem 2.56. Mary read a book in four days. The number of pages she read on the second day was two times less than the number of pages she read on the first day, the number of pages she read on the fourth day was 50% of the number of pages she read on the second day. Given that the ratio of the number of pages she read on the third day to the number of pages she read on the fourth day is 3 : 4. What percent of the book did Mary read on the first day?

(A) $\dfrac{20}{3}$ (B) 40   (C) 70   (D) $\dfrac{1600}{31}$ (E) 30.3

Problem 2.57. Let $x_1, x_2, ..., x_{20}$ be a sequence of numbers. Given that $x_1 = 2$ and

$$x_{n+1} = \frac{n+2}{n+1} x_n,$$

for $n = 1, 2, ..., 19$. What is the value of x_{20}?

(A) 21   (B) 20   (C) $\dfrac{21}{20}$ (D) $\dfrac{20}{21}$ (E) $\dfrac{21}{16}$

Problem 2.58. Let x and y be different numbers, such that $20x + 13x^2 = 20y + 13y^2$. What is the value of $x + y$?

(A) $\dfrac{13}{20}$ (B) $-\dfrac{20}{13}$ (C) $-\dfrac{13}{20}$ (D) $\dfrac{20}{13}$ (E) 0

Problem 2.59. *Let a straight line containing incenter I of $\triangle ABC$ be parallel to side AC. Let M, N be intersection points of this line with sides AB, BC, respectively. Given that $MN = 10$ and $AC = 15$. What is the value of the perimeter of trapezoid $AMNC$?*

(A) 35 (B) 40 (C) 30 (D) 50 (E) 55

Problem 2.60. *Let a, b, c, d be digits. What is the greatest possible positive value of integer n satisying the following equation?*

$$\frac{1}{n} = 0.\overline{abcd} = 0.abcdabcd....$$

(A) 1111 (B) 909 (C) 303 (D) 9999 (E) 3333

Problem 2.61. *Let AB and CD be the bases of trapezoid $ABCD$ (see the figure). Let M, N and P, K be points on bases AB and CD, respectively, such that $AB + CD = 18$ and*

$$Area(AMPD) = Area(MNKP) = Area(NBCK).$$

What is the length of the midsegment of trapezoid $MNKP$?

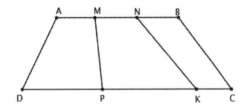

(A) 2.5 (B) 4.5 (C) 6 (D) 9 (E) 3

Problem 2.62. *Let the angle measures of a triangle with sides a, b, c form an arithmetic sequence. Given that $a \leq b \leq c$ and that $a, \sqrt{2}b, 3c$ form a geometric sequence. What is the value of the common ratio of this geometric sequence?*

(A) $\sqrt{2}$ (B) $\sqrt{3}$ (C) $\sqrt{6}$ (D) 2 (E) 3

Problem 2.63. *Let $ABCD$ be a convex quadrilateral of area 18 and with vertex coordinates $A(1,1), B(2,7), C(m,n), D(6,3)$ on the xy-coordinate plane. What is the value of $m + n$?*

(A) 12 (B) 7 (C) 8 (D) 11 (E) 10

Problem 2.64. *Given that $2^{\sqrt[4]{\log_2 3}} = 3^{\sqrt[4]{\log_3^3 x}}$. What is the value of x?*

(A) $\sqrt{2}$ (B) 2 (C) $\sqrt[3]{2}$ (D) 4 (E) 3

Problem 2.65. *The principal of a school wants to give three identical math books, two identical physics books, and two identical chemistry books as gifts to the best four 12th graders. Each of the students should receive at least one of the books, but not two identical ones. In how many different ways can the principal give out the books?*

(A) 108 (B) 144 (C) 72 (D) 120 (E) 100

Problem 2.66. *There are three types of solutions with 10%, 20%, 30% concentrations of salt, respectively. Given that if we mix the first and the second types, then the mixture has 16% concentration of salt. Given also that if we mix the second and the third types, then the mixture has 24% concentration of salt. What concentration of salt would the mixture of the first and the third types have?*

(A) 15 (B) 25 (C) 20 (D) 5 (E) 40

Problem 2.67. *On the first day of his work a painter painted $\frac{1}{3}$ part of the wall. On the second day, he painted $\frac{1}{5}$ of the remaining part of the wall, and so on. On the nth day, he painted $\frac{1}{2n+1}$ of the remaining part of the wall. What part of the wall the painter still needed to paint after the sixth day?*

(A) $\frac{1}{3}$ (B) $\frac{1}{125}$ (C) $\frac{512}{3003}$ (D) $\frac{1024}{3003}$ (E) $\frac{2048}{3003}$

Problem 2.68. *The centers of eight congruent spheres of radii 1 are vertices of a cube with an edge length of 3. What is the radius of a sphere that is externally tangent to the eight given spheres?*

(A) $1.5\sqrt{3}$ (B) 1 (C) $1.5\sqrt{3} - 1$ (D) $1.5\sqrt{3} + 1$ (E) 3

Problem 2.69. *Let a circle passing through vertices A, C of triangle ABC intersects sides AB, BC at points M, N, respectively. Given that the side lengths of triangle BMN are integers and $AB = 8, BC = 6$. What is the greatest possible length of line segment MN?*

(A) 5 (B) 4 (C) 8 (D) 7 (E) 6

Problem 2.70. *Consider the set $M = \{-14, -13, ..., -5, 1, 2, ..., 9\}$. How many ordered triples (a, b, c) exist, such that $a + b + c = 0$, where a, b, c are elements of M?*

(A) 201 (B) 195 (C) 192 (D) 222 (E) 102

Problem 2.71. *What is the value of the integer part of the following expression?*

$$\log_8 9 + \log_9 10 + ... + \log_{63} 64.$$

(A) 56 (B) 60 (C) 57 (D) 59 (E) 53

Problem 2.72. *Let M be the set of all six-digit numbers, such that all digits of a number are from the set $\{1, 3, 4, 5, 6, 7, 9\}$. What is the probability that a randomly chosen number from set M is divisible by 7?*

(A) $\frac{1}{2}$ (B) $\frac{1}{3}$ (C) $\frac{1}{7}$ (D) $\frac{1}{5}$ (E) $\frac{1}{4}$

Problem 2.73. *Given two congruent squares with a side length of 4, such that their centers coincide (see the figure). What is the smallest possible value of the area of the common interior part of these squares?*

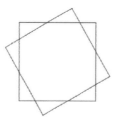

(A) 8 (B) $32(\sqrt{2} - 1)$ (C) $8\sqrt{2}$ (D) $16(\sqrt{2} - 1)$ (E) $4(\sqrt{2} + 1)$

Problem 2.74. *Four random vertices of a regular dodecagon (twelve-sided polygon) are chosen (see the figure). What is the probability that these four vertices form a quadrilateral with exactly two right angles?*

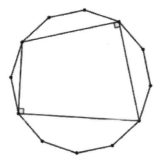

(A) $\dfrac{1}{4}$ (B) $\dfrac{1}{3}$ (C) $\dfrac{1}{6}$ (D) $\dfrac{8}{33}$ (E) $\dfrac{10}{33}$

Problem 2.75. *How many real numbers a exist, such that the equation $z^3 + iz^2 + 5z + a = 0$ has a root of the form $z = x + iy$, where x and y are integers?*

(A) 0 (B) 2 (C) 4 (D) 3 (E) 1

2.4 AMC 12 type practice test 4

Problem 2.76. What is the value of the expression $10 - (9 - (8 - (7 - (6 - (5 - (4 - (3 - (2 - 1))))))))$?

(A) 1 (B) 3 (C) 0 (D) 7 (E) 5

Problem 2.77. It took 5 hours to drive a car from city A to city B. In the first hour the car covered 50 miles and starting from the second hour it covered 10 miles per hour more than in each previous hour. What is the total distance (in miles) covered by the car?

(A) 350 (B) 260 (C) 300 (D) 340 (E) 320

Problem 2.78. The area of the intersection part of two circles is 10% of the area of one of the circles, and 40% of the area of the other circle. What is the ratio of the radius of the larger circle to the radius of the smaller circle?

(A) $\sqrt{2}$ (B) 4 (C) 2 (D) π (E) $\sqrt{3}$

Problem 2.79. What is the range of the solutions of the following inequality?

$$\frac{x}{2^x} + \sqrt[5]{x} + \frac{|x|}{x} \geq 0.$$

(A) $[0, +\infty)$ (B) $(0, +\infty)$ (C) $(-\infty, 0)$ (D) $[1, +\infty)$ (E) $(-\infty, +\infty)$

Problem 2.80. David solved 93 problems over the course of several days. Each day he solved either 3, 4 or 5 problems. Given that the total number of days during which he solved 3 or 4 problems is not more than 12, and the total number of days during which he solved 4 or 5 problems is not more than 17. What is the greatest number of days he could have spent solving 4 problems in a day?

(A) 4 (B) 6 (C) 8 (D) 7 (E) 5

Problem 2.81. A palindrome number is a number that remains the same when its digits are reversed, for example 16061. What is the smallest possible positive difference of two different four-digit palindrome numbers?

(A) 10 (B) 11 (C) 9 (D) 110 (E) 99

Problem 2.82. Given three spheres, the radii of the first two spheres are 3 and 4. The sum of the total surface area of these two spheres is equal to the total surface area of the third sphere. What is the value of the volume of the third sphere?

(A) 125π (B) 200π (C) 100π (D) $\dfrac{400\pi}{3}$ (E) $\dfrac{500\pi}{3}$

Problem 2.83. Let the longest side of triangle ABC be AC. Let BH be an altitude, such that $AC = 4 \cdot BH$ and $\angle C = 15°$. What is the value of the angle measure (in degrees) of $\angle A$?

(A) 15 (B) 30 (C) 75 (D) 45 (E) 60

Problem 2.84. *Let the edge length of a cube be 4 inches. A right circular cylinder and a square prism are removed from this cube (see the figure). Given that the radius of the base of the cylinder is 1 inch and the centers of its bases coincide with the centers of two opposite faces of the cube. The centers of the bases of the square prism coincide with the centers of two other opposite faces of the cube. The base edge length of the square prism is 2 inches and all its lateral faces are parallel to the corresponding faces of the cube. What is the value of the volume (in cubic inches) of the remaining solid?*

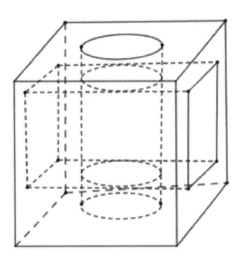

(A) $56 - 4\pi$ (B) $64 - 4\pi$ (C) 48   (D) 32   (E) $48 - 2\pi$

Problem 2.85. *Let p and q be given numbers, such that $p - 2q, p - q, p + q$ and 40 form a geometric sequence. What is the smallest possible value of the sum of the first three terms of this sequence?*

(A) 120   (B) 35   (C) 20   (D) 24   (E) 30

Problem 2.86. *What is the value of the greatest solution of the equation $3^{x^2+x} = 4^{x+1}$?*

(A) -1 (B) 0   (C) $2\log_3 2$ (D) $\log_4 3$ (E) $\log_3 2$

Problem 2.87. *Each of the inhabitants of TruLi island $(A, B, C, D$ and $E)$ either always tells the truth or always lies. They once said the following:*
A said: "C is a truth-teller."
B said: "D is a truth-teller."
C said: "E is a truth-teller."
D said: "B is a truth-teller."
E said: "B is a liar."
What is the value of the product of the number of truth-tellers and the number of liars on TruLi island?

(A) 6   (B) 4   (C) 0   (D) 5   (E) 3

Problem 2.88. *What is the greatest possible value of a, such that the following system of equations has a solution in the set of real numbers?*

$$\begin{cases} y = x^2 + 7x + 3a, \\ x = y^2 + 7y + \dfrac{1}{3}a^3. \end{cases}$$

(A) 9   (B) 6   (C) 7   (D) 3   (E) 2

Problem 2.89. Let D be a point on side AC of triangle ABC. Given that $AB = 6, AC = 9, CD = 5$ and $BD = 8$. What is the value of the length of side BC?

(A) $5\dfrac{1}{3}$ (B) 10 (C) 8 (D) 12 (E) 14

Problem 2.90. Let the probability that it will rain some day during a given week be equal to $\dfrac{1}{3}$. Which one of the following events has the greatest probability to happen?

(A) It will rain exactly 2 days during a week.
(B) It will rain exactly 3 days during a week.
(C) It will rain exactly 4 days during a week.
(D) It will rain exactly 5 days during a week.
(E) It will rain exactly 6 days during a week.

Problem 2.91. Ben randomly chooses a number from each of the sets: $\{1, 3, 5, 7, 9, 11\}$ and $\{2, 4, 6, 8, 10\}$. What is the probability that the sum of the chosen numbers is a multiple of 3?

(A) $\dfrac{4}{15}$ (B) $\dfrac{1}{3}$ (C) $\dfrac{1}{5}$ (D) $\dfrac{1}{4}$ (E) $\dfrac{1}{6}$

Problem 2.92. Let $ABCDEF$ be a convex hexagon, such that $\angle B = \angle D = \angle F = 120°, \angle C = 2 \cdot \angle ACE$ and $\angle A = 2 \cdot \angle CAE$. Given that the area of triangle ACE is equal to $10\sqrt{3}$ and the area of hexagon $ABCDEF$ is equal to $11\sqrt{3}$. What is the value of $|DE - EF|$?

(A) $\sqrt{3}$ (B) 3 (C) $3\sqrt{3}$ (D) 6 (E) $6\sqrt{3}$

Problem 2.93. The interior of a square with vertices (3, 1), (5, 3), (3, 5), (1,3) is cut out of the interior of a square with vertices (0, 0), (0, 6), (6, 6), (6, 0). An ant departs from (0, 0) and moves in the remaining interior of the bigger square, such that from each point (i, j) it moves straight either to the point $(i + 1, j)$ or to the point $(i, j + 1)$, where i and j are integers. Given that the ant stops at (6, 6). How many possible paths are there for the ant to reach (6, 6) from (0, 0)?

(A) 51 (B) 102 (C) 124 (D) 50 (E) 200

Problem 2.94. Sam is randomly choosing n unit squares from different columns of $n \times n$ square grid. The probability that n chosen unit squares are in different columns and in different rows is equal to $\dfrac{5}{324}$. What is the value of n?

(A) 5 (B) 7 (C) 8 (D) 6 (E) 12

Problem 2.95. Let n be a positive integer and $(a_n), (b_n)$ be increasing arithmetic sequences with integer terms. Given that $a_1 = b_1 = 1$ and k is such a positive integer greater than 2 that $\dfrac{1}{a_k} + \dfrac{1}{b_k} = \dfrac{1}{10}$. What is the value of $\dfrac{1}{a_2} + \dfrac{1}{b_2}$?

(A) $\dfrac{1}{6}$ (B) $\dfrac{1}{3}$ (C) $\dfrac{1}{4}$ (D) $\dfrac{1}{2}$ (E) 1

Problem 2.96. Let $p(x)$ be a polynomial with integer coefficients. Given that $|p(x)| \le 2010x^2$ for all real values of x. How many such polynomials exist?

(A) 4021 (B) 2010 (C) 3 (D) 2011 (E) 100

Problem 2.97. Let x be a real number. What is the smallest possible value of the following expression?

$$|x - \log_2 3| + |x - \log_3 4| + |x - \log_4 5|.$$

(A) 0 (B) 1 (C) $\log_2 \dfrac{3\sqrt{5}}{5}$ (D) $\log_2 3$ (E) 2

Problem 2.98. What is the sum of the last two digits of the sum $1^3 + 2^3 + ... + 2010^3$?

(A) 10 (B) 7 (C) 9 (D) 5 (E) 17

Problem 2.99. What is the value of n for which the solution of the following inequality represents a union of n disjoint intervals?

$$\log_x(2x) \cdot \log_{3x}(4x) \cdot \log_{5x}(6x) \cdot \log_{7x}(8x) \cdot \log_{9x}(10x) < 0.$$

(A) 10 (B) 4 (C) 6 (D) 3 (E) 5

Problem 2.100. Let M be the set $\{1, 2, ..., 16\}$. What is the total number of all three-element subsets of M, such that each of the subsets does not contain consecutive numbers?

(A) 378 (B) 350 (C) 364 (D) 320 (E) 560

2.5 AMC 12 type practice test 5

Problem 2.101. *A car left city A toward city B at 8:30 AM. The driver reached city B, rested for 25 minutes in city B and drove back to city A, reaching it at 1:05 PM. How long (in minutes) was the car driving on the road?*

(A) 275   (B) 225   (C) 240   (D) 250   (E) 295

Problem 2.102. *Given that* $1 - \dfrac{1}{1 - \dfrac{1}{1+x}} = 2$. *What is the value of x?*

(A) -1 (B) 0 (C) $\dfrac{1}{3}$ (D) -0.5 (E) 0.5

Problem 2.103. *Five numbers are inserted between integers 1 and 2, such that all seven numbers together form an arithmetic sequence. What is the value of the smallest inserted number?*

(A) $\dfrac{6}{5}$ (B) $\dfrac{7}{6}$ (C) $\dfrac{4}{3}$ (D) $\dfrac{3}{2}$ (E) 1.1

Problem 2.104. *There is $500 in each of three envelopes. The first envelope contains only $10 bills, the second one only $20 bills, and the third one only $50 bills. One, two, and three bills are taken out of these envelopes (in any order). What is the smallest possible amount (in dollars) that could have been taken out?*

(A) 130   (B) 120   (C) 110   (D) 230   (E) 100

Problem 2.105. *If the width of a rectangle is increased by 2 units and the length is decreased by 2 units, then the area is equal to 28sq. units. On the other hand, if the width of the initial rectangle is decreased by 2 units and the length is increased by 2 units, then the area is equal to 24sq. units. What is the value (in sq. units) of the area of the initial rectangle?*

(A) 28   (B) 30   (C) 24   (D) 16   (E) 18

Problem 2.106. *Let m, n be integers, such that $2^m \cdot 3^n = a$ and $2^n \cdot 3^m = b$. What is the value of $3^{n^2 - m^2}$?*

(A) $a^n \cdot b^m$ (B) $a^m \cdot b^{-n}$ (C) $(ab)^{n-m}$ (D) $a^{-n} \cdot b^{-m}$ (E) $a^n \cdot b^{-m}$

Problem 2.107. *Let (b_n) be a geometric sequence with common ratio equal to 7. Given that the sum of its first five terms is equal to 41. What is the value of $b_6 + b_7 - b_1 - b_2$?*

(A) 1968   (B) 2009   (C) 41   (D) 287   (E) 100

Problem 2.108. *Given five unit squares (see the figure). What is the length of the side of the greatest square?*

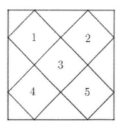

(A) 2   (B) 3   (C) $2\sqrt{2}$ (D) $\sqrt{7}$ (E) $\sqrt{6}$

Problem 2.109. Let $f(x) = ax^2 + bx + c$ and $f(x + 2030) = x^2 - x - 1$, for any value of x. What is the value of $b^2 - 4ac$?

(A) 2030 (B) -3 (C) 5 (D) 4 (E) 25

Problem 2.110. Let ABC be a triangle, such that $AB = 1$ and $BC = 12$. Given that the length (in units) of median BM is an integer number. What is the value of the length of BM?

(A) 3 (B) 4 (C) 7 (D) 5 (E) 6

Problem 2.111. What is the value of the following expression?

$$1^3 + 2 \cdot 2^2 + 2 \cdot 3^2 + ... + 2 \cdot 19^2 + 20^2 + 1 \cdot 2 + 2 \cdot 3 + 3 \cdot 4 + ... + 19 \cdot 20.$$

(A) 7999 (B) 8000 (C) 7200 (D) 789 (E) 800

Problem 2.112. How many positive integers have the following property: the integer is 6 less than the sum of the squares of all of its digits?

(A) 1 (B) 3 (C) 2 (D) 0 (E) 6

Problem 2.113. Let ABC and $A_1B_1C_1$ be triangles. Given that $AB = A_1B_1 = 20$, $BC = B_1C_1 = 10\sqrt{2}$ and $\angle ABC - \angle A_1B_1C_1 = 90°$. What is the greatest possible value of $AC - A_1C_1$?

(A) $10\sqrt{2}$ (B) 10 (C) 20 (D) 30 (E) $20\sqrt{2}$

Problem 2.114. On the $xy-$coordinate plane points $(0,0)$, $(0,3)$, $(4,3)$ form a triangle. For what value of k does the line $y = kx + 1$ divide the perimeter of this triangle in two equal halfs?

(A) $\dfrac{4}{3}$ (B) $\dfrac{3}{4}$ (C) 2 (D) $-\dfrac{1}{2}$ (E) $\dfrac{1}{2}$

Problem 2.115. Let a, b be real numbers, such that $i + 2i^2 + ... + 2009i^{2009} = a + bi$. What is the value of $a + b$?

(A) 1 (B) 2008 (C) 2010 (D) 2009 (E) -1

Problem 2.116. Let two circles of radii 2 and 4 be externally tangent to each other. Given that line l is tangent to both circles and that both circles lie on the same side of line l. What is the radius of the circle that is externally tangent to both circles and with the center on line l?

(A) $\dfrac{8}{7}$ (B) 1 (C) 1.5 (D) $\dfrac{7}{8}$ (E) $\dfrac{2}{3}$

Problem 2.117. Let the sum of all the terms of the infinite geometric sequence $1, \dfrac{1}{\sqrt[2009]{2}}, ...$ be equal to S. What is the value of the sum of the infinite geometric sequence $1, \dfrac{1}{S}, ...$?

(A) $\sqrt[2009]{4}$ (B) $\sqrt[2009]{2} + 1$ (C) 2 (D) $1 + \dfrac{1}{\sqrt[2009]{2}}$ (E) $\sqrt[2009]{2}$

Problem 2.118. Let n be a positive integer. How many terms of the sequence $x_n = 10^n - 3^n + 2^n + 5$ are perfect squares?

(A) 0 (B) 2 (C) 1 (D) 4 (E) 3

Problem 2.119. *What is the value of the greatest possible perimeter of a pentagon for which there exists an annulus with the area of 4π, such that given pentagon can be entirely insterted into that annulus? Note that some (or all) vertices of given pentagon can also lie on the circumferences of two concentric circles of that annulus. (An annulus is a plane figure bounded by the circumferences of two concentric circles with different radii).*

(A) 10π (B) 20   (C) 10   (D) 20π (E) 15

Problem 2.120. *Let the diagonals of a convex quadrilateral ABCD intersect at point O. Given that the areas of $\triangle ABO$ and $\triangle CDO$ are equal to 8 and 18, respectively. What is the smallest possible value of the area of ABCD?*

(A) 39   (B) 50   (C) 52   (D) 40   (E) $10\sqrt{5}$

Problem 2.121. *Let $P(x) = x^3 + ax^2 + bx + c$ be a polynomial with real coefficients. Given that both equations $P(x) = 3$ and $P(x) = 1$ have two different real roots. What is the value of $2a^3 - 9ab + 27c$?*

(A) 25   (B) 37   (C) 42   (D) 54   (E) 63

Problem 2.122. *Let all 8 edges of a regular square pyramid are equal to 1 unit. Given that a plane parallel to one of its lateral faces intersects the pyramid and the perimeter of the intersection part is equal to $2\frac{1}{3}$. What is the value (in sq. units) of the area of the intersection part?*

(A) $\dfrac{5\sqrt{3}}{36}$ (B) $\dfrac{\sqrt{3}}{8}$ (C) $\dfrac{\sqrt{3}}{16}$ (D) $\dfrac{4\sqrt{3}}{81}$ (E) $\dfrac{1}{64}$

Problem 2.123. *Let point (1, 2) be the vertex of the graph of the first quadratic function and point (3, 6) be the vertex of the graph of the second quadratic function. The graph of the second function passes through the vertex of the graph of the first function and the graph of the first function passes through the vertex of the graph of the second function. What is the length of the line segment formed by two intersection points of the x-axis with the graph of the second function?*

(A) 6   (B) 2   (C) $2\sqrt{6}$ (D) $\sqrt{6}$ (E) $2\sqrt{5}$

Problem 2.124. *Let n be a positive integer, denote by $A(n)$ be the number of digits of 2^n, by $B(n)$ be the number of digits of $2^{A(n)}$ and by $C(n)$ the number of digits of $2^{n+A(n)}$. How many elements are in the range of the function $A(n) + B(n) - C(n)$?*

(A) 1   (B) 5   (C) 2009   (D) 2010   (E) 2

Problem 2.125. *Let (x_n) be a number sequence defined as follows: $x_1 = 1, x_2 = \dfrac{1}{2}, x_3 = \dfrac{1}{2}$ and $x_{n+3} = 2x_{n+2}x_{n+1} - x_n$. How many three-digit numbers n are there, such that $x_n = 1$?*

(A) 0   (B) 75   (C) 900   (D) 100   (E) 76

2.6 AMC 12 type practice test 6

Problem 2.126. *A train left city A at 8 : 00 AM and arrived in city B after 45 minutes. It stopped in city B for 10 minutes and continued on to city C. The train covered the distance between city B and city C in 75 minutes. At what time did the train arrive in city C?*

(A) 11:00 AM (B) 10:05 AM (C) 10:00 AM (D) 10:10 AM (E) 10:30 AM

Problem 2.127. *What is the opposite of the value of the following expression $\frac{1}{3} - \frac{3}{4}$?*

(A) -2.4 (B) $-\frac{5}{12}$ (C) $\frac{5}{12}$ (D) 1 (E) $\frac{1}{2}$

Problem 2.128. *Six painters working at the same constant rate can completely paint an apartment in 8 hours. How many painters were working if it took 12 hours to paint the apartment?*

(A) 5 (B) 3 (C) 4 (D) 2 (E) 1

Problem 2.129. *What is the value of the following expression?*

$$\frac{1^2 + 1 \cdot 2 + 2^2}{1^3 \cdot 2^3} + \frac{2^2 + 2 \cdot 3 + 3^2}{2^3 \cdot 3^3} + \ldots + \frac{9^2 + 9 \cdot 10 + 10^2}{9^3 \cdot 10^3}.$$

(A) 1 (B) 0.9 (C) 0.99 (D) 0.5 (E) 0.999

Problem 2.130. *For how many positive values of x is $\frac{20}{x+1} + \frac{1}{3(x+1)}$ a positive integer?*

(A) 0 (B) 10 (C) 19 (D) 20 (E) 25

Problem 2.131. *A shop bought a coat for \$500. The shop inteded to sell the coat for some price, but sold it 5% more than the intended sales price. Given that for that deal the shop generated a total profit of \$67. What was the intended sales price of the coat?*

(A) \$560 (B) \$540 (C) \$550 (D) \$530 (E) \$520

Problem 2.132. *A fisherman makes a round-trip and takes his boat 10 miles into a lake from its shore and back (covering the same distance). The average speed of the boat is 2.5 miles per hour. The fisherman catches (in average) 0.5 pounds of fish per hour. How many pounds of fish does the fisherman catch per one round-trip?*

(A) 4 (B) 3 (C) 2 (D) 5 (E) 6

Problem 2.133. *Given a rectangular prism, such that its volume is equal to $\sqrt{3}$. Each edge of the rectangular prism is increased $\sqrt{3}$ times. What is the volume of the new rectangular prism?*

(A) $3\sqrt{3}$ (B) 9 (C) 3 (D) 27 (E) $9\sqrt{3}$

Problem 2.134. *The width and length of the frame of a painting are in the proportion 4:5. The respective dimensions of the painting are in the proportion 3:4 and the painting is the same distance from the frame on each side. What is the ratio of the area of the frame to the area of the painting?*

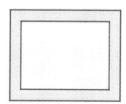

(A) $\dfrac{1}{2}$ (B) $\dfrac{2}{3}$ (C) $\dfrac{1}{3}$ (D) $\dfrac{1}{4}$ (E) $\dfrac{1}{10}$

Problem 2.135. *One combine harvester can harvest a field on its own in 12 hours. A second combine harvester can harvest the same field on its own in 18 hours. How many hours will it take both combine harvesters, working together, to harvest the whole field, if they also take a 48 minute break?*

(A) 7.2   (B) 7   (C) 9   (D) 8   (E) 10

Problem 2.136. *All faces of four identical $1 \times 1 \times 1$ cubes are numbered from 1 to 6. Moreover, the opposite faces are numbered with numbers 1 and 2, 3 and 4, 5 and 6. These four cubes are used to form a $1 \times 1 \times 4$ rectangular prism (placed on the table), such that the sum of 17 numbers on the visible faces is the smallest possible. What is the the value of the sum of these 17 numbers?*

(A) 44   (B) 40   (C) 45   (D) 42   (E) 74

Problem 2.137. *Let function $f(x)$ be defined on $[0,1]$. Given that the sum of the greatest and the smallest values of $f(x)$ is equal to 10. What is the value of the sum of the greatest and the smallest values of function $g(x) = 13 - 2 \cdot f(1-x)$?*

(A) -7 (B) 7   (C) 26   (D) 13   (E) 6

Problem 2.138. *Let AB be the diameter of a semicircle with center O. Semicircles, with diameters OA and OB, are drawn. Circle S is drawn, such that it is tangent to these three semicircles (see the figure). Given that the radius of circle S is equal to r. What is the value of $\dfrac{AB}{r}$?*

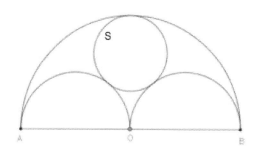

(A) 5   (B) 4   (C) 6   (D) 4.5   (E) 3

Problem 2.139. *Let x, y be integers. What is the total number of all points $M(x,y)$, such that the following inequality holds true?*
$$|4x - 24| + |3y + 15| \leq 12.$$

(A) 27   (B) 32   (C) 25   (D) 29   (E) 28

Problem 2.140. *What is the units digit of the following number?*
$$2008^{2007^{2008}} + 2007^{2008^{2007}}.$$

(A) 1   (B) 3   (C) 5   (D) 7   (E) 9

Problem 2.141. *Let a,b be numbers greater than 1, such that the following three numbers*
$$\log(a^2b), \log(a^2b^2), \log(a^5b^2)$$
form a geometric sequence. What is the value of the common ratio of this geometric sequence?

(A) $\dfrac{2}{3}$ (B) 1.5   (C) 2   (D) 0.5   (E) 3

Problem 2.142. *Let n be a positive integer and (a_n) be a sequence defined as follows: $a_1 = 2430$ and*
$$a_n = \begin{cases} \dfrac{a_{n-1}}{3}, & \text{if } 3 \mid a_{n-1}, \\ 2a_{n-1} + 1, & \text{if } 3 \nmid a_{n-1}, \end{cases}$$
for $n = 2, 3, \ldots$. How many numbers among the terms $a_1, a_2, \ldots, a_{2008}$ are divisible by 3?

(A) 662   (B) 104   (C) 6   (D) 9   (E) 7

Problem 2.143. *Let $SABC$ be a pyramid, such that $AB = AC = 13, BC = 5$ and $\angle ASB = \angle ASC = 90°, \angle BSC = 60°$. What is the value of the volume of $SABC$?*

(A) 50   (B) 25   (C) $25\sqrt{3}$ (D) $75\sqrt{3}$ (E) $50\sqrt{3}$

Problem 2.144. *What is the coefficient of x^{23} in the expansion of the following expression?*
$$(24 + 23 \cdot x + 22 \cdot x^2 + \ldots + 1 \cdot x^{23})(1 + x + x^2 + \ldots + x^9)(1 + x + x^2 + \ldots + x^{11}).$$

(A) 1320   (B) 132   (C) 529   (D) 1000   (E) 5290

Problem 2.145. *Let CM be a median in a right triangle with legs AC and BC. Given that $AC = 6$ and $BC = 8$. What is the value of the sum of the inradii of triangles ACM and BCM?*

(A) $2\dfrac{1}{3}$ (B) 3.5   (C) $2\dfrac{5}{6}$ (D) 3   (E) 2

Problem 2.146. *How many five-digit numbers are there which are divisible by 11 and are composed (without repetitions) only of digits 1,2,3,4,8?*

(A) 10   (B) 12   (C) 24   (D) 11   (E) 0

Problem 2.147. *Let a square be constructed externally on each side of a regular hexagon with a side length of 1. What is the radius of the circle which passes through all the vertices of these squares that are not the vertices of the hexagon?*

(A) $\sqrt{3}$ (B) 2   (C) $\dfrac{\sqrt{6}+\sqrt{2}}{2}$ (D) $\sqrt{5}+1$ (E) $\sqrt{2}$

Problem 2.148. *Given that the solutions of the following equation are the vertices of a quadrilateral on the complex plane.*
$$(z+i)^4 + (z-i)^4 = 16.$$
What is the value of the perimeter of that quadrilateral?

(A) 20   (B) 8   (C) $2\sqrt{5}$ (D) $8\sqrt{5}$ (E) $8\sqrt{2}$

Problem 2.149. *Let D be a given point on side BC of $\triangle ABC$. Given that $CD = 1, BD = 3$ and $\angle C = \dfrac{\pi}{3}$. What is the greatest possible value of the measure of $\angle BAD$?*

(A) $\dfrac{3\pi}{8}$ (B) $\dfrac{\pi}{6}$ (C) $\dfrac{\pi}{2}$ (D) $\dfrac{\pi}{3}$ (E) $\dfrac{\pi}{4}$

Problem 2.150. *Let n be a positive integer and $(a_n), (b_n)$ be sequences of complex numbers defined as follows: $a_1 = i, b_1 = 1$ and*
$$a_{n+1} = \sqrt[3]{2}a_n - b_n,$$
$$b_{n+1} = a_n + \sqrt[3]{2}b_n,$$
where $n = 1, 2, \dots$. What is the value of the expression $\left|\dfrac{a_{2008}}{b_{2008}}\right|$?

(A) $\sqrt[3]{2}$ (B) 1   (C) 2   (D) $\dfrac{1}{2}$ (E) $\sqrt[3]{4}$

2.7 AMC 12 type practice test 7

Problem 2.151. *The area of a park is 300 sq.m and the area of a nearby park is 430 sq.m. The area of the first park is p% less than the area of the second park. What is the value of the closest integer to p?*

(A) 29 (B) 30 (C) 31 (D) 32 (E) 33

Problem 2.152. *Let a positive integer a be 25% less than a positive integer b. Given that their sum is equal to 49. What is the value of a?*

(A) 21 (B) 30 (C) 27 (D) 14 (E) 35

Problem 2.153. *What is the smallest possible perimeter of a triangle with different integer side lengths?*

(A) 5 (B) 6 (C) 12 (D) 8 (E) 9

Problem 2.154. *How many positive odd divisors does 30^{10} have?*

(A) 121 (B) 1331 (C) 665 (D) 666 (E) 667

Problem 2.155. *Given two congruent circles on the plane with non-coincident centers. How many of the transformations below map one circle onto another?*
- *parallel translation.*
- *point symmetry.*
- *line symmetry.*
- *rotation by 30° angle.*

(A) 4 (B) 3 (C) 2 (D) 1 (E) 0

Problem 2.156. *What is the value of the greatest integer that cannot be written as a sum of two composite numbers?*

(A) 101 (B) 11 (C) 111 (D) 2019 (E) 42

Problem 2.157. *Let ABC be a triangle, such that $AC = 10, BC = 17$ and $AB = 21$. Let M and N be points on side AB, such that $CM^2 = AM \cdot BM$ and $CN^2 = AN \cdot BN$ (see the figure). What is the value of the area of triangle CMN?*

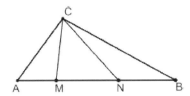

(A) 34 (B) 42 (C) 30 (D) 36 (E) $10\sqrt{3}$

Problem 2.158. *A geometric sequence consists of five terms. The arithmetic mean of the first four terms of this sequence is equal to 10. The arithmetic mean of the last four terms is equal to 30. What is the fifth term of this sequence?*

(A) 40 (B) 52 (C) 64 (D) 72 (E) 81

Problem 2.159. *Consider four congruent circles on a plane, such that they are pairwise non-concentric and they divide the plane in n parts. How many values of n are possible?*

(A) 10 (B) 9 (C) 8 (D) 7 (E) 6

Problem 2.160. Let x and y be randomly chosen numbers from $[0, 1]$. What is the probability that the following inequality holds true?
$$|x - 0.5| + |y - 1.5| \leq 1.$$

(A) $\dfrac{1}{4}$ (B) $\dfrac{3}{4}$ (C) $\dfrac{1}{6}$ (D) $\dfrac{5}{6}$ (E) $\dfrac{1}{2}$

Problem 2.161. Given 22 circles, such that 21 of them have radius 1 (see the figure). Given also that any two circles that have a common point are pairwise tangent. What is the value of the circumference of the largest circle?

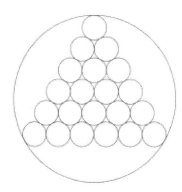

(A) $\dfrac{2\pi}{3}(10\sqrt{3} + 3)$ (B) $\dfrac{20}{3}\pi$ (C) $\dfrac{20\sqrt{3}}{3}\pi$ (D) 13π (E) 20π

Problem 2.162. Let x and y be positive numbers, such that $x \neq 1$ and $y \neq 1$. Given that $\log_4 x = \log_y 3$ and $\log_3 y = \log_2 \dfrac{32}{x}$. What is the value of the expression $\log_x 2 + \log_y 3$?

(A) 2   (B) 3   (C) 2.5   (D) 3.5   (E) 4

Problem 2.163. Let a and b be complex numbers, such that for all values of x it holds true:
$$(x^2 - 2x + 3)(x^2 - 4x + 12) = (x^2 + ax + 6)(x^2 + bx + 6).$$

What is the value of the expression $|a - b|$?

(A) $2\sqrt{2}$ (B) $3\sqrt{2}$ (C) $6\sqrt{2}$ (D) 12   (E) 2

Problem 2.164. Let n be a positive integer. Consider a number sequence (a_n), such that $a_1 = 1, a_2 = 5$ and
$$a_{n+2} = 5a_{n+1} - 6a_n,$$
where $n = 1, 2, \ldots$. What is the value of the expression $(a_{100} - 3 \cdot a_{99})$?

(A) 2^{99} (B) 2^{100} (C) 3^{99} (D) 3^{100} (E) $3^{100} + 2^{99}$

Problem 2.165. Let a, b, c be positive numbers, such that $\log_2 a - 2 = \log_3 b = \log_5 c$ and $a + b = c$. What is the value of the product $a \cdot b \cdot c$?

(A) 1000   (B) 3600   (C) 100   (D) 625   (E) 4900

Problem 2.166. *How many eight-digit numbers are there which are divisible by 11 and are composed (without repetitions) only of digits 1, 2,..., 8?*

(A) 3456 (B) 1152 (C) 5040 (D) 40320 (E) 4608

Problem 2.167. *Let 2×10 rectangular grid be randomly covered by ten 1×2 rectangles (dominos). What is the probability that two cells in the fifth column from the left are covered by different dominos?*

(A) $\dfrac{64}{89}$ (B) $\dfrac{49}{89}$ (C) $\dfrac{25}{89}$ (D) $\dfrac{40}{89}$ (E) $\dfrac{1}{2}$

Problem 2.168. *Let the vertices of $\triangle ABC$ lie on the surface of a sphere with a radius of 13 (see the figure). Given that $AB = 5$ and $m\angle ACB = 30°$. What is the value of the distance from the center of the sphere to the plane containing triangle ABC?*

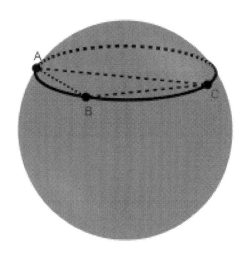

(A) 10 (B) 11 (C) 12 (D) 6 (E) 8

Problem 2.169. *Let $f(x)$ be a monic polynomial function of degree five, such that*

$$f(2021) = -2, f(2022) = 1, f(2023) = 2, f(2024) = 1, f(2025) = -2.$$

What is the value of $f(2020)$?

(A) -5 (B) -12 (C) -127 (D) -210 (E) -2100

Problem 2.170. *Let n be a positive integer and (a_n) be a number sequence defined as follows: $a_1 = a_2 = 1, a_3 = \dfrac{1}{2}$ and*

$$a_{n+1} = \frac{a_n^3 a_{n-2}}{a_{n-1}^3 + a_{n-2}a_{n-1}a_n},$$

where $n = 3, 4, \ldots$. What is the value of the following expression?

$$\frac{a_1}{a_2} + 2\frac{a_2}{a_3} + 3\frac{a_3}{a_4} + \ldots + 10\frac{a_{10}}{a_{11}}.$$

(A) 11! (B) 11!-1 (C) 10!+3 (D) 10!+11 (E) $10 \cdot 11$

Problem 2.171. Let $z = \cos\dfrac{2\pi}{31} + i\sin\dfrac{2\pi}{31}$. What is the value of the following expression?

$$(1-z+z^2)(1-z^2+z^4)(1-z^4+z^8)(1-z^8+z^{16})(1-z^{16}+z^{32}).$$

(A) 1   (B) 0   (C) $\dfrac{1}{2}$ (D) $\dfrac{\sqrt{2}}{2}$ (E) 2^{31}

Problem 2.172. A league of soccer teams participate in a tournament. Any two teams play with each other exactly once. By the end of the tournament, exactly n games end in a draw and the total points gained in the tournament are equal to 2019. What is the value of the sum of all possible values of n? (Note: A winning team gets 3 points, a losing team gets 0 points, while a draw is 1 point for each team.)

(A) 2019   (B) 2050   (C) 2150   (D) 4038   (E) 4080

Problem 2.173. Let vertices A, B, C of triangle ABC lie on circles with center O and radii $\sqrt{52}$, 3 and 5, respectively. Given that point O lies in the interior of triangle ABC and $\angle ABC = 120°$, $\angle BAC = 30°$. What is the value of the length of side AB?

(A) 4   (B) 5   (C) 6   (D) 7   (E) 8

Problem 2.174. Let n be a positive integer and (x_n) be a number sequence defined as follows: $x_1 = \dfrac{1}{\sqrt[6]{2}}$ and

$$\dfrac{x_n}{x_{n-1}^2} = 2^{1/2^n},$$

where $n = 2, 3, \ldots$. What is the value of the expression $(x_{2019})^{3 \cdot 2^{2019}}$?

(A) 2^{2019} (B) 2^{10} (C) 0.5   (D) 0.25   (E) 3

Problem 2.175. Let n be a positive integer, such that $\dfrac{(n!)^2}{(n+3)!}$ is also a positive integer. What is the smallest possible number of divisors of the expression $(n+1)(n+2)(n+3)$?

(A) 24   (B) 30   (C) 36   (D) 12   (E) 20

2.8 AMC 12 type practice test 8

Problem 2.176. Some of the students in the class are students in an after-school math circle as well. Most of the students in the class participated in the AMC 12 test. It appeared that $\frac{7}{8}$ of the students atteding the math circle (from this class) and $\frac{5}{6}$ of the students not attending the math circle (from this class) participated in the AMC 12. Given that the total number of students in the class is not more than 19. What part of the class participated in the AMC 12?

(A) $\frac{1}{2}$ (B) $\frac{2}{3}$ (C) $\frac{6}{7}$ (D) $\frac{3}{4}$ (E) $\frac{8}{9}$

Problem 2.177. Let points $A(-1, 4)$ and $B(2, 10)$ lie on a rectangular coordinate plane. Given that point $C(x_0, y_0)$ lies on segment AB, such that $AC : CB = 1 : 2$. What is the value of $x_0 + y_0$?

(A) 6 (B) 5 (C) 4 (D) -6 (E) -5

Problem 2.178. Given that the mean of the numbers $30, 31, ..., 29 + 2n$ is equal to the mean of the numbers $100, 101, ..., 99 + n$. What is the value of the sum of all digits of n?

(A) 10 (B) 5 (C) 8 (D) 7 (E) 6

Problem 2.179. Given that the numbers $n!, (n+1)!, n!(n+19)$ form an arithmetic sequence, where n is a positive integer. What is the value of the product of the digits of n?

(A) 1 (B) 24 (C) 30 (D) 3 (E) 8

Problem 2.180. Let m and n be positive integers, such that both the sum and the difference of numbers $mn + m + n - 53$ and $9n - 6m + 3$ are prime numbers. What is the value of the sum of all digits of the product mn?

(A) 7 (B) 9 (C) 10 (D) 13 (E) 15

Problem 2.181. Let the lengths of two of the sides of triangle ABC be 6, 8 and its area be 24. What is the value of the perimeter of triangle ABC?

(A) 20 (B) 24 (C) 22 (D) 27 (E) 26

Problem 2.182. Let $f(x) = 2x^3 - 3x^2 + x + 5$ and

$$A = f\left(\frac{1}{2020}\right) + f\left(\frac{2}{2020}\right) + f\left(\frac{3}{2020}\right) + ... + f\left(\frac{2018}{2020}\right) + f\left(\frac{2019}{2020}\right).$$

What is the value of the sum of all digits of A?

(A) 1 (B) 10 (C) 15 (D) 18 (E) 22

Problem 2.183. For how many positive integers n the values of fractions $\frac{20n + 19}{n + 817}$ and $\frac{19n + 61}{n + 817}$ are positive integers too?

(A) 0 (B) 1 (C) 3 (D) 5 (E) 11

Problem 2.184. What is the smallest possible integer value of x, such that $\log_2 x, \log_4 x, \log_2 \frac{x}{4}$ are the lengths of sides of the same triangle?

(A) 5 (B) 10 (C) 16 (D) 17 (E) 18

Problem 2.185. *A hotel with infinite number of rooms has room numbers labeled 1, 2, 3,... Given that a guest is assigned to room i with probability $\frac{1}{2^i}$. What is the probability that the positive difference of the room numbers assigned to two random guests is 2?*

(A) $\frac{1}{3}$ (B) $\frac{2}{3}$ (C) $\frac{1}{12}$ (D) $\frac{1}{6}$ (E) $\frac{1}{4}$

Problem 2.186. *What is the total number of all unordered pairs of edges of a regular octahedron, such that no plane can pass through them?*

(A) 12   (B) 24   (C) 36   (D) 42   (E) 48

Problem 2.187. *A circle σ of radius $2 - \sqrt{3}$ and line l tangent to circle σ at point B lie in the same plane. Let A be a point outside circle σ and AM be tangent to circle σ, such that it does not intersect with line l. Given that $AM = 1$. Let C, D be points on line l, such that $BC = BD = 1$. What is the angle measure (in degrees) of $\angle CAD$?*

(A) 15   (B) 30   (C) 45   (D) 60   (E) 75

Problem 2.188. *Let a be the smallest possible value of x, such that the values of the functions $y = \frac{5}{6}x + \frac{1}{3}$ and $y = \frac{25}{16}x - \frac{1}{2}$ are positive integers. What is the value of the sum of all digits of a^3?*

(A) 1   (B) 2   (C) 3   (D) 6   (E) 8

Problem 2.189. *Let point $M(x_0, y_0)$ and circle σ lie in the same plane. Let MB and MD be secants, such that rays BM and DM intersect σ at points A and C, respectively. Given that $AB = CD$ and $A(2,4), B(3,6), C(0,2)$. What is the value of $4x_0 + y_0$?*

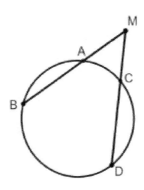

(A) 2   (B) 3   (C) 8   (D) 0   (E) 5

Problem 2.190. *What is the probability that a random positive divisor of 2020^5 is not ending in 0?*

(A) $\frac{8}{33}$ (B) $\frac{2}{11}$ (C) $\frac{2}{3}$ (D) $\frac{3}{4}$ (E) $\frac{10}{101}$

Problem 2.191. *What is the total number of all complex numbers z, such that $0, z, z^2$ are the vertices of an isosceles right triangle?*

(A) 0   (B) 2   (C) 4   (D) 6   (E) 8

Problem 2.192. Let ABC be a triangle, such that $AC = 3, BC = 5, AB = 7$ and its each side is the diameter of a semicircle (see the figure). Let S_1, S_2, S_3 be the areas of the grey parts. What is the value of $S_1 + S_3 - S_2$?

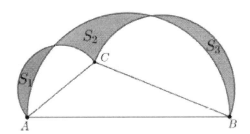

(A) $\dfrac{15}{8}(2\sqrt{3} - \pi)$ (B) $\dfrac{15}{4}(2\sqrt{3} - \pi)$ (C) $\dfrac{\pi}{16} - \dfrac{\sqrt{3}}{18}$ (D) 4 (E) $\pi + 2$

Problem 2.193. Let A be the smallest positive ten-digit number that is divisible by 36 and whose base 10 representation consists of only 4's and 9's, with at least one of each. What is the remainder of A after division by 11?

(A) 1 (B) 3 (C) 5 (D) 8 (E) 10

Problem 2.194. Let the base of pyramid $SABCD$ be a square with side length 4. Given that $SA = 3$, where SA is the altitude to the base. M is a point inside the pyramid equidistant from all its faces. Denote by h the distance from point M to any of its faces. What is the value of h?

(A) 1 (B) 2 (C) 3 (D) 4 (E) 5

Problem 2.195. For how many quadratic trinomials $x^2 + px + q$ each of the numbers $p + q + 1$ and $p - q + 1$ is a root of given quadratic trinomial?

(A) 0 (B) 1 (C) 4 (D) 2 (E) 3

Problem 2.196. A random four-digit number is chosen. What is the probability that the digits of the chosen number are four consecutive numbers?

(A) $\dfrac{1}{30}$ (B) $\dfrac{9}{500}$ (C) $\dfrac{7}{300}$ (D) $\dfrac{11}{450}$ (E) $\dfrac{7}{80}$

Problem 2.197. Let $ABCD$ be a convex quadrilateral, such that $AD = 1, CD = 2, \angle ABC = 90°$ and $AB : BC = 3 : 4$. What is the greatest possible value of the area of $ABCD$?

(A) $\dfrac{30 + \sqrt{1224}}{25}$ (B) 2 (C) $\dfrac{30 + \sqrt{1201}}{25}$ (D) 2.5 (E) 3

Problem 2.198. How many ten-digit numbers $\overline{a_1a_2...a_{10}}$ with nonzero digits exist, such that each of the three-digit numbers $\overline{a_1a_2a_3}, \overline{a_2a_3a_4}, ..., \overline{a_8a_9a_{10}}$ is not divisible by 3?

(A) $2^8 \cdot 3^{10}$ (B) $2^8 \cdot 3^{12}$ (C) $2^{10} \cdot 3^{10}$ (D) $2^{12} \cdot 3^{12}$ (E) $2^{12} \cdot 3^8$

Problem 2.199. Let n be a positive integer and sequence (x_n) be defined as follows: $x_1 = 1$ and $x_{n+1} = x_n + \dfrac{1}{x_n}$, where $n = 1, 2, ...$. What is the value of the sum of all digits of the smallest possible value of n, such that $x_n > 8$?

(A) 5 (B) 12 (C) 16 (D) 21 (E) 24

Problem 2.200. Points $O(0, 0), A(1, 1), B(2, 3), C(-1, 2), D(-2, 5)$ are drawn on a rectangular coordinate plane. Let M, N be any points on line segments AB, CD, respectively. What is the area of a figure formed by the locus of all points X, such that $\overrightarrow{OX} = \overrightarrow{OM} + \overrightarrow{ON}$?

(A) $3 + 2\sqrt{2}$ (B) 6 (C) 4.5 (D) $3\sqrt{3}$ (E) 5

2.9 AMC 12 type practice test 9

Problem 2.201. *What is the value of the following expression?*
$$(2^{-1} + 3^{-1} - 4^{-1})^{-1}.$$

(A) $\frac{7}{12}$ (B) 12 (C) $\frac{12}{7}$ (D) 1 (E) 9

Problem 2.202. *David and Ann have together some candies. Assume that David gives half of his candies to Ann, afterward Ann gives half of her candies to David, then David gives half of his candies to Ann, and finally Ann gives half of her candies to David. In the end, David and Ann have 15 and 9 candies, respectively. How many candies did Ann have in the beginning?*

(A) 24 (B) 20 (C) 16 (D) 8 (E) 10

Problem 2.203. *A factory that produces cars overperformed the plan for January by 20% and underperformed the plan for February by 25%. It turned out, that in total the factory exactly performed their two months plan. By how many percents is the plan for February less than the plan for January?*

(A) 25 (B) 45 (C) 20 (D) 10 (E) 5

Problem 2.204. *Given that the sum of two consecutive odd prime numbers is not divisible by 4. What is the smallest possible number of divisors that this sum can have?*

(A) 8 (B) 9 (C) 4 (D) 5 (E) 6

Problem 2.205. *There were some candies on the table. Alexa ate half of the candies and a half of one candy. Afterward, her brother ate half of the remaining candies and a half of one candy. Finally, there is one candy left on the table. How many candies were there on the table in the beginning?*

(A) 15 (B) 9 (C) 13 (D) 11 (E) 7

Problem 2.206. *What is the total number of negative solutions of the following equation?*
$$(1+x)(1+x^2)(1+x^4) = 1.$$

(A) 5 (B) 4 (C) 3 (D) 0 (E) 1

Problem 2.207. *In triangle ABC points D, E are on sides AB, BC, respectively. Let circumcenter O of triangle ABC be the intersection point of segments CD and AE. Given that $BD = BO = BE$ and $\angle ABC = n°$. What is the value of n?*

(A) 72 (B) 60 (C) 30 (D) 36 (E) 45

Problem 2.208. *John has cut with a knife (by straight lines) each of the identical chocolate bars into two pieces, afterward he has cut (by straight lines) some of the pieces into two pieces and finished his actions. Given that every time John performed the cutting action he selected either exactly one chocolate bar to cut or exactly one piece to cut. Given also that John performed the cutting action 19 times and in the end he had 28 pieces. What is the total number of chocolate bars?*

(A) 7 (B) 4 (C) 14 (D) 9 (E) 8

Problem 2.209. *Let x, y be real numbers, such that $\sqrt{x} - \sqrt{x - 2019} = \sqrt{y + 2019} - \sqrt{y}$. What is the value of $x - y$?*

(A) 1 (B) 100 (C) 2019 (D) -2019 (E) 1001

Problem 2.210. Let each of the areas of congruent equilateral triangles ABC and DEF be equal to 16 (see the figure). Given that
$$AE + AK = EB + BC + CK.$$
What is the value of the area of triangle AEK?

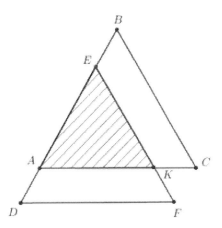

(A) 5   (B) 4   (C) 8   (D) 9   (E) 10

Problem 2.211. Let $O(0)$ and $A(i)$ be points on the complex plane. Let point $B(z)$ passes to point C after a rotation by $90°$ around point O, point C passes to point D after a rotation by $90°$ around point A and point D passes to point B after a rotation by $90°$ around point O. What is the value of z?

(A) 0   (B) $1+i$ (C) $-i$ (D) 1   (E) i

Problem 2.212. Given numbers 1, 2, 3, 4, 5, 6, 7, 8, 9, 10. Given that Ann chooses one of those numbers, then Emily chooses one of the remaining numbers, afterward David chooses one of the remaining numbers. What is the probability that Ann's number is less than Emily's number and Emily's number is less than David's number?

(A) $\dfrac{1}{2}$ (B) $\dfrac{2}{3}$ (C) $\dfrac{1}{6}$ (D) $\dfrac{1}{12}$ (E) $\dfrac{1}{24}$

Problem 2.213. Given four positive integers, such that one of them is one-digit number, one is a two-digit number, one is a three-digit number and one is a four-digit number. Given that the sum of two-digit and four-digit numbers is more by 2 than the sum of one-digit and three-digit numbers. What is the value of the sum of one-digit and three-digit numbers?

(A) 1008   (B) 11100   (C) 900   (D) 801   (E) 716

Problem 2.214. What is the value of the sum of all solutions of the following equation?
$$6^x - 4 \cdot 3^x - 27 \cdot 2^x + 108 = 0.$$

(A) 1   (B) 4   (C) 5   (D) 6   (E) 2

Problem 2.215. Let a_1, a_2, a_3, a_4, a_5 be five pairwise different positive integers, such that a_1, a_2, a_3, a_4, a_5 is a geometric sequence. Given that the number of positive divisors of each of the numbers a_1 and a_5 is 5. What is the number of positive divisors of a_4?

(A) 5   (B) 4   (C) 6   (D) 7   (E) 8

Problem 2.216. Let sides AD, BC of a tangential quadrilateral $ABCD$ be tangent to its incircle at points M, N, respectively. Let line segments MN and AC intersect at point E. Given that $AM = 3, CN = 6, AC = 18$. What is the value of the length of line segment AE?

(A) 12 (B) 6 (C) 8 (D) 4.5 (E) 9

Problem 2.217. Let two cars start moving simultaneously toward each other from places A and B. Given that cars move with constant speeds. Given also that when the cars cross each other the first car coming from place A continues its movement with speed equal to 1.25 times its initial speed, and the second car continues its movement with speed equal to 1.8 times its initial speed. Afterward, it turned out that the first car reached place B at the moment when the second car reached place A. What is the value of the ratio of the initial speed of the first car to the initial speed of the second car?

(A) 2 (B) 1.2 (C) 2.5 (D) 3 (E) 4

Problem 2.218. From all positive integers not including digit 1 in their decimal notation was chosen 100th smallest number. What is the value of the sum of all digits of the chosen number?

(A) 7 (B) 6 (C) 9 (D) 3 (E) 8

Problem 2.219. Let M, N be points on bases BC, AD of trapezoid $ABCD$, respectively. Given that $ABMN$ and $DCMN$ are tangential quadrelaterals and $MN = 10$. What is the value of the distance between the incenters of $ABMN$ and $DCMN$?

(A) 5 (B) 11 (C) 10 (D) 20 (E) 12

Problem 2.220. The positive integer a is called "interesting number", if the value of the expression $\dfrac{20!}{a}$ is a perfect square. What is the number of all digits of the smallest "interesting number"?

(A) 4 (B) 5 (C) 3 (D) 6 (E) 12

Problem 2.221. Given a triangular pyramid $SABC$ with the volume V. Let V_1 be the volume of the triangular pyramid with the vertices at the intersection points of the medians of the faces of given pyramid. What is the value of $\dfrac{V}{V_1}$?

(A) 27 (B) 9 (C) $\dfrac{9}{4}$ (D) 3 (E) 4

Problem 2.222. What is the greatest possible value of the following expression?

$$\sin x + \sin y - \sin x \sin y.$$

(A) 3 (B) 2 (C) 1 (D) $\sqrt{3}$ (E) $\sqrt{2}$

Problem 2.223. What is the total number of all nine-digit numbers $\overline{a_1...a_9}$ with pairwise different non-zero digits, such that four numbers among the following eight two-digit numbers $\overline{a_1a_2}, \overline{a_2a_3}, ..., \overline{a_8a_9}$ are divisible by 6?

(A) 120 (B) 240 (C) 480 (D) 1680 (E) 3360

Problem 2.224. Let $p(x)$ and $q(x)$ be polynomials with integer coefficients, such that

$$p^2(x) - p(x)q(x) + q^2(x) = x^2 - x + 1$$

holds true for any x. What is the total number of all such pairs $(p(x), q(x))$?

(A) 0 (B) 12 (C) 8 (D) 10 (E) 16

Problem 2.225. Let a two-digit number be chosen randomly. What is the probability that the chosen number can be represented as a difference of cubes of two positive integers?

(A) $\dfrac{1}{2}$ (B) $\dfrac{1}{10}$ (C) $\dfrac{4}{45}$ (D) $\dfrac{1}{15}$ (E) $\dfrac{2}{3}$

2.10 AMC 12 type practice test 10

Problem 2.226. *What is the value of the following expression?*
$$\frac{(-1)^{-1}}{(-3)^{-3}} + \frac{1}{(-2)^{-2}}.$$

(A) 31 (B) 23 (C) -23 (D) $\frac{31}{108}$ (E) -31

Problem 2.227. *James is repairing torn pages in each of 4 books. He spends the same amount of time for the reparation of each book and after the reparation of each book he rests for 3 minutes. Given that James started to fix the first book at 14 : 00 and finished the third book at 15 : 06. At what time was James done with reparation of all 4 books?*

(A) 15 : 28 (B) 15 : 26 (C) 15 : 29 (D) 15 : 30 (E) 16 : 00

Problem 2.228. *An archer shot several arrows and each time he got either 5 or 7 points. All together, the archer received 58 points. What is the total number of shots?*

(A) 11 (B) 10 (C) 12 (D) 8 (E) 6

Problem 2.229. *Jack, Marta, Rand, Todd and 6 more students went to pick mushrooms. Every two of these 10 pupils picked a different number of mushrooms and all together they picked 45 mushrooms. Jack and Marta together picked 16, Marta and Rand together picked 15, Rand and Todd together picked 13 mushrooms. How many mushrooms did Todd pick?*

(A) 10 (B) 7 (C) 5 (D) 6 (E) 0

Problem 2.230. *A tourist group consists of French and Chinese tourists, such that none of the group members has dual citizenship. Given that 15 % of French tourists and 10% of Chinese tourists (in this group) speak english. What is the smallest possible number of tourists in the group?*

(A) 15 (B) 35 (C) 20 (D) 25 (E) 30

Problem 2.231. *Let the columns of 10×10 square grid be numbered from left to right by numbers from 1 to 10, and the rows be numbered from bottom to top by numbers from 1 to 10. Let the intersection point of ith column and j^{th} row be number $ij + i + j$. What percent of all numbers in the square are even numbers?*

(A) 75 (B) 50 (C) 30 (D) 25 (E) 35

Problem 2.232. *The regular n-gon is divided into several regular m-gons, where $n > m$. What is the value of $m + n$?*

(A) 9 (B) 6 (C) 7 (D) 8 (E) 10

Problem 2.233. *What is the value of the following expression*
$$\frac{2^{\log_2^2 3}}{3^{\log_2 12}}.$$

(A) 1 (B) $\frac{1}{9}$ (C) 9 (D) 3 (E) 2

Problem 2.234. *There are 3 black and 4 white balls in a bag. Mary and Ann one after the other each took a single ball. Mary starts and the girls look at the color of the ball after taking it. Whoever takes a white ball first is the winner. What is the probability that Ann wins?*

(A) $\dfrac{24}{35}$ (B) $\dfrac{1}{2}$ (C) $\dfrac{11}{35}$ (D) $\dfrac{1}{3}$ (E) $\dfrac{2}{3}$

Problem 2.235. *Let m and n be the number of all pairwise non-congruent triangles with integer sides, such that each of them has a perimeter of 2017 and 2018, respectively. What is the value of $n - m$?*

(A) 0 (B) 168 (C) -168 (D) 5 (E) -5

Problem 2.236. *What is the area of a triangle with sides lying on lines given by the following equations $x = 4, y = 3$ and $x + y = 11$?*

(A) 16 (B) 28 (C) 14 (D) 20 (E) 8

Problem 2.237. *Given that pairwise different numbers a, b, c, d form an arithmetic sequence and their sum is equal to 20. What is the sum of all solutions of the following equation?*

$$(x - a)(x - b)(x - c) + (x - b)(x - c)(x - d) = 0.$$

(A) 20 (B) 10 (C) 12 (D) 15 (E) 40

Problem 2.238. *Let ABCD be a cyclic quadrilateral, where $\angle BAC + \angle ADB = 60°$ and $AB = 7, BC = 8$. What is the value of the length of AC?*

(A) 8 (B) 10 (C) 11 (D) 13 (E) 12

Problem 2.239. *Given an equilateral triangle with side length of 6 and a circle intersecting the triangle, such that its diameter is one of the sides of this triangle. What is the area of the part of the circle that is inside of the triangle?*

(A) $\dfrac{3(\pi + 3\sqrt{3})}{2}$ (B) $\dfrac{27\sqrt{3}}{4}$ (C) $\dfrac{9\pi}{2}$ (D) $\dfrac{3(2\pi + 3\sqrt{3})}{2}$ (E) $\dfrac{27}{2}$

Problem 2.240. *Let 1×1 cell be randomly removed from 7×7 square grid. What is the probability that (after 1×1 cell was removed) the remaining part of 7×7 square grid can be covered by sixteen 1×3 rectangular grids?*

(A) $\dfrac{1}{2}$ (B) $\dfrac{9}{49}$ (C) $\dfrac{8}{49}$ (D) $\dfrac{16}{49}$ (E) $\dfrac{3}{4}$

Problem 2.241. *What is the value of the volume of a quadrilateral pyramid, such that its all eight edges have a length of 6?*

(A) 72 (B) $72\sqrt{2}$ (C) $36\sqrt{2}$ (D) 36 (E) $48\sqrt{2}$

Problem 2.242. *Jack takes without looking n balls out of a box that contains white, red and blue balls. The probability of taking out a white ball is $\dfrac{1}{2}$, the probabilities of taking out red and blue balls are $\dfrac{1}{3}$ and $\dfrac{1}{6}$, respectively. Given that the probability that "out of n balls exactly 2 balls are white and exactly 3 balls are red" is equal to the probability that "out of n balls exactly 3 balls are white and exactly 3 balls are red." What is the value of n?*

(A) 10 (B) 9 (C) 8 (D) 7 (E) 6

Problem 2.243. *A positive integer is called a "special number" if it has at least three different prime divisors. Let n be a "special number", we denote by $Q(n)$ the sum of the pairwise products of all distinct prime divisors of n, for example*

$$Q(30) = 2 \cdot 3 + 3 \cdot 5 + 2 \cdot 5 = 31,$$

$$Q(60) = 2 \cdot 3 + 3 \cdot 5 + 2 \cdot 5 = 31.$$

What is the smallest possible even value of $Q(n)$?

(A) 236 (B) 32 (C) 64 (D) 100 (E) 320

Problem 2.244. *Let ABC be a right triangle, such that $\angle C = 90°$. Given that regular hexagons $AC_1C_2C_3C_4C$ and $AB_1B_2B_3B_4B$ are constructed externally on the sides of triangle ABC, such that points C_2, C_3, B_2, B_3 lie on one circle. What is the value (in degrees) of $\angle B$?*

(A) 30 (B) 45 (C) 15 (D) 60 (E) 75

Problem 2.245. *Let function f be defined as follows $f : \{0, 1, 2, 3, ...\} \times \{0, 1, 2, 3, ...\} \to \{0, 1, 2, 3, ...\}$, $f(0, 0) = 2$ and*

$$f(i,j) = \begin{cases} 2f(i-1,j) - 1, & \text{if } j = 0, i \geq 1, \\ 3f(i,j-1) - 2, & \text{if } i = 0, j \geq 1, \\ f(i-1,j) + f(i,j-1) - f(i-1,j-1), & \text{if } i \geq 1, j \geq 1. \end{cases}$$

What is the value of $f(3, 4)$?

(A) 7 (B) 89 (C) 25 (D) 91 (E) 100

Problem 2.246. *A three-digit number is called "ordinary", if the positive difference of its some two neighbor digits is greater than or equal to 2. What is the total number of "ordinary" three-digit numbers?*

(A) 823 (B) 825 (C) 650 (D) 640 (E) 700

Problem 2.247. *In a meeting 32 chairs are placed around a circular table and each of them is occupied by one person. After a break, each of them can occupy the eighth chair (in any direction) counting from the chair they have occupied before the break, where the counting starts from an adjacent chair (after the break also each chair is occupied by one person). After the break, in how many ways can these people be seated around the table?*

(A) 65536 (B) 128 (C) 32 (D) 2^{32} (E) 16

Problem 2.248. *Given a rectangular box with dimensions $a \times b \times c$, where a, b, c are positive integers and $a \leq b \leq c$. The volume and the sum of the lengths of all edges of this rectangular box are numerically equal. How many such ordered triples (a, b, c) are possible?*

(A) 10 (B) 6 (C) 14 (D) 11 (E) 8

Problem 2.249. *Let ABC be an obtuse non-isosceles triangle. Let M and N be points on the longest side AC, such that $AM = CN = 39$ and $AB : BM = BC : BN = 8 : 5$. Given that the lenght of the altitude of triangle ABC drawn to side AC is 24. What is the length of AC?*

(A) 40 (B) 36 (C) 100 (D) 64 (E) 128

Problem 2.250. *Let polynomial $p(x)$ be given as follows:*

$$p(x) = 1 + 2x + 3x^2 + ... + 2015x^{2014}.$$

What is the value of $\left| p\left(e^{i\frac{\pi}{3}}\right) \right|$?

(A) 0 (B) 2014 (C) 2016 (D) 1 (E) 2015

2.11 AMC 12 type practice test 11

Problem 2.251. What is the value of the following expression

$$\sqrt[6]{2^5 \cdot \sqrt[7]{2^6 \cdot \sqrt[8]{2^7 \cdot \sqrt[9]{2^8 \cdot \sqrt[10]{2^{10}}}}}}.$$

(A) $\sqrt[6]{2}$ (B) 2   (C) 4   (D) 8   (E) 1024

Problem 2.252. Let the initial speed of a car be 70 miles per hour. At first its speed was increased by 10%, then the obtained speed was decreased by 10 miles per hour. What is the value (in miles per hour) of the final speed of the car?

(A) 70   (B) 69   (C) 68   (D) 67   (E) 50

Problem 2.253. What is the value of the product of all solutions of the following equation?

$$x\left(x^2 - 4 + \frac{1}{x}\right) = 1.$$

(A) 0   (B) -4   (C) 4   (D) 1   (E) -1

Problem 2.254. Let a be a positive integer, such that the difference of the arithmetic mean and geometric mean of numbers a and 9 is equal to 8. What is the sum of all digits of a?

(A) 13   (B) 14   (C) 15   (D) 20   (E) 21

Problem 2.255. Let CD be the bisector of angle ACB in triangle ABC. Given that $AC = CD$ and $\angle ACB = 108°$. Let $\angle A = n \cdot \angle B$. What is the value of n?

(A) 3   (B) 5   (C) 6   (D) 7   (E) 10

Problem 2.256. As a homework a student needs to solve algebra and geometry problems, all together 27 problems. Given that the student has managed to solve 80% of algebra problems and 75% of geometry problems. What part of the entire homework has the student finished?

(A) $\frac{7}{9}$ (B) $\frac{2}{3}$ (C) $\frac{5}{9}$ (D) $\frac{1}{2}$ (E) $\frac{3}{7}$

Problem 2.257. What is the total number of all integers a, such that the following equation has exactly two real solutions?

$$|2^x - 10| = a.$$

(A) 21   (B) 19   (C) 11   (D) 10   (E) 9

Problem 2.258. Let in all cells of 3×3 square grid be written positive integers (one number per cell), such that the sum of all written numbers is equal to 215 and their product is equal to 2020. How many 1 is written in the cells of 3×3 square grid?

(A) 3   (B) 4   (C) 5   (D) 6   (E) 7

Problem 2.259. Given that an interior angle of a regular n-gon is equal to $k°$, where k is a positive integer. What is the total number of all possible values of n?

(A) 12   (B) 15   (C) 18   (D) 20   (E) 22

Problem 2.260. *From ten consecutive positive integers were chosen seven numbers, such that their sum is equal to 2020. What is the greatest possible value of the sum of three not chosen numbers?*

(A) 600 (B) 700 (C) 875 (D) 880 (E) 890

Problem 2.261. *A rectangular prism is called "beautiful", if its three dimensions are positive integers. Let a rectangular prism M be divided by three planes parallel to its faces into eight "beautiful" rectangular prisms. Given that the volumes of four of these "beautiful" rectangular prisms are equal to 1, 2, 3, 5. What is the value of the total surface area of rectangular prism M?*

(A) 108 (B) 84 (C) 72 (D) 60 (E) 30

Problem 2.262. *Let 2 two-digit numbers be chosen at random. What is the probability of the event that the positive difference of these 2 two-digit numbers is also a two-digit number?*

(A) $\dfrac{36}{89}$ (B) $\dfrac{2}{5}$ (C) $\dfrac{72}{89}$ (D) $\dfrac{4}{5}$ (E) $\dfrac{9}{10}$

Problem 2.263. *Let the graph of function $y = g(x)$ be symmetric to the graph of function $y = x^2$ with respect to point $M = (3, 9)$. What is the value of the length of the line segment, such that its endpoints are the intersection points of the graph of function $y = g(x)$ with the x–axis?*

(A) 2 (B) 6 (C) $6\sqrt{2}$ (D) $6\sqrt{3}$ (E) 18

Problem 2.264. *What is the total number of all three-digit numbers, not containing any zero digit, for which there is a digit such that after erasing that digit the obtained two-digit number is divisible by 3? For example three-digit numbers 121 and 123 satisfy these conditions.*

(A) 120 (B) 159 (C) 729 (D) 540 (E) 513

Problem 2.265. *What is the value of the product of all solutions of the following equation?*

$$3^{\sqrt{\log_3 x}} = 4^{\sqrt{\log_2^3 x}}.$$

(A) 1 (B) $2^{\sqrt{\log_2 \sqrt[4]{3}}}$ (C) 2 (D) 3 (E) 4

Problem 2.266. *Let $ABCD$ be a rhombus, such that $\angle A = 45°$ and $AC = 13$. Assume that ray BD intersects the circumcircle of triangle ABC at point E. What is the value of the length of line segment DE?*

(A) 12 (B) 13 (C) $13\sqrt{2}$ (D) 20 (E) 26

Problem 2.267. *Given that*
$$2009 = m + n + k,$$
where m, n, k are positive integers. At most with how many zeros can the product mnk end with?

(A) 5 (B) 6 (C) 7 (D) 8 (E) 10

Problem 2.268. *Let*
$$p(x) = x^4 + x^3 - 7x^2 - x + 6,$$
and $q(x)$ is a polynomial with real coefficients, such that $p(x) + q(x) \geq 0$ and $p(x)q(x) \leq 0$ for any value of x. What is the value of the expression $|q(-2) + q(4)|$.

(A) 0 (B) 200 (C) 198 (D) 10 (E) 12

Problem 2.269. *A committee of eight scientists consists of local scientists and visitor scientists. Every scientist brought two identical copies of a self-authored manuscript. Each of them exchanged the first copy of a self-authored manuscript with one copy of a self-authored manuscript of other attendee scientist and the second copy of a self-authored manuscript with one copy of self-authored manuscript of another attendee scientist. Given that any two scientists can exchange a copy of their self-authored manuscripts if one of them is a local scientist and another one is a visitor scientist. In how many different ways can they perform such exchanges?*

(A) 72 (B) 75 (C) 84 (D) 88 (E) 90

Problem 2.270. *Let u and v be positive numbers, such that $|u - v| \geq 1$. What is the smallest possible value of the expression $uv + \frac{u}{v} + \frac{v}{u}$?*

(A) 3 (B) 3.5 (C) 4 (D) $\sqrt{35}$ (E) 6

Problem 2.271. *Consider a coordinate system O_{xyz}. Let solid φ consists of all points $M(x, y, z)$, such that for each of them the following inequality holds true:*

$$x^2 + y^2 + z^2 \leq |x| + |y| + |z|.$$

What is the value of the volume of solid φ?

(A) $\dfrac{\sqrt{3}\pi}{2}$ (B) $2\sqrt{3}\pi + 4$ (C) $\sqrt{3}\pi + 4$ (D) $\dfrac{4}{3}\pi$ (E) 1

Problem 2.272. *Given that*

$$\sum_{n=1}^{\infty} \frac{\cos(n\phi)}{2^n} = \sum_{n=1}^{\infty} \frac{\sin(n\phi)}{2^n}.$$

What is the value of $\sin(2\phi)$?

(A) 1 (B) $\dfrac{3}{4}$ (C) $\dfrac{1}{2}$ (D) $\dfrac{1}{4}$ (E) 0

Problem 2.273. *Given point M on side AB and point N on side BC of triangle ABC, such that*

$$\frac{BM}{CN} = \frac{BN}{AM} = \frac{3}{5},$$

and

$$\angle BMN - \angle BNM = 60°.$$

What is the value of $\dfrac{MN}{AC}$?

(A) $\dfrac{3}{5}$ (B) $\dfrac{2}{5}$ (C) $\dfrac{3}{7}$ (D) $\dfrac{5}{7}$ (E) $\dfrac{1}{2}$

Problem 2.274. *Let m and n be positive integers, such that $n \mid (m + 20)$ and $m \mid (n + 21)$. What is the value of the sum of all digits of the greatest possible value of $m + n$?*

(A) 3 (B) 4 (C) 5 (D) 10 (E) 12

Problem 2.275. *Let the inradius of triangle ABC be equal to 1. Given that $\angle BAC = 30°$ and $\angle ABC = 45°$. A random segment of length 1 is drawn inside of triangle ABC, such that its projections on sides BC, AC, AB are a, b, c, respectively. What is the probability of the event that*

$$\sqrt{2}a + 2b = (\sqrt{3} + 1)c?$$

(A) $\dfrac{1}{3}$ (B) $\dfrac{5}{12}$ (C) $\dfrac{2}{3}$ (D) $\dfrac{7}{12}$ (E) $\dfrac{3}{4}$

2.12 AMC 12 type practice test 12

Problem 2.276. What is the value of the sum of the opposite of 2020 and the reciprocal of $\dfrac{1}{2019}$?

(A) $-\dfrac{1}{2019 \cdot 2020}$ (B) $2019\dfrac{1}{2020}$ (C) $-2020\dfrac{1}{2019}$ (D) -1 (E) 1

Problem 2.277. Boat tour ticket costs 25$ for an adult and 10$ for a child. Given that a group of 7 people paid 130$ for 7 tickets. How many children are in this group?

(A) 1 (B) 3 (C) 4 (D) 6 (E) 7

Problem 2.278. Let one of the numbers 1,2,...,9 be equal to the arithmetic mean of the other eight numbers. What is the value of the sum of all digits of the sum of the other eight numbers?

(A) 2 (B) 3 (C) 4 (D) 6 (E) 11

Problem 2.279. Let n be any positive integer. What is the number of elements of the set

$$\left\{1, \frac{1}{2}, \frac{1}{3}, ..., \frac{1}{n}, ...\right\},$$

such that each of them is a solution of the following inequality?

$$100x^2 - 25x + 1 < 0.$$

(A) 13 (B) 14 (C) 15 (D) 16 (E) 19

Problem 2.280. The first shop sells 450 grams of some type of candy for 5$ and the second shop sells 500 grams of the same type of candy for 6$. How much more in percentage is the price of the candy in the second shop compared to the first shop?

(A) 12 (B) 10 (C) 9 (D) 8 (E) 5

Problem 2.281. For which value of m do the graphs of functions $y = x - 1$, $y = 3x - 5$, $y = mx - 41$ intersect at one point?

(A) 5 (B) 10 (C) 19 (D) 21 (E) 22

Problem 2.282. What is the value of the sum of all real solutions of the following equation?

$$6^x - 2 \cdot 3^x - 27 \cdot 2^x + 54 = 0.$$

(A) 2 (B) 3 (C) 3.5 (D) 4 (E) 6

Problem 2.283. What is the value of the greatest possible solution of the following equation?

$$\log_2 x = \sqrt{\log_3 x}.$$

(A) 1 (B) $2^{\log_3 2}$ (C) 2 (D) 3 (E) 3.2

Problem 2.284. Given three pairwise different positive integers, such that the sum of each two of them is greater than the consecutive number of the third number. What is the smallest possible value of the sum of these three numbers?

(A) 10 (B) 11 (C) 12 (D) 13 (E) 14

Problem 2.285. What is the value of the value of the following expression?

$$\frac{3}{1\cdot 2} - \frac{5}{2\cdot 3} + \frac{7}{3\cdot 4} - \frac{9}{4\cdot 5} + \frac{11}{5\cdot 6} - \frac{13}{6\cdot 7} + \frac{15}{7\cdot 8} - \frac{17}{8\cdot 9} + \frac{19}{9\cdot 10}.$$

(A) 1.1 (B) 0.9 (C) 0.75 (D) 0.6 (E) 0.5

Problem 2.286. Let a, b, c, d be such digits that $\overline{ab}, \overline{cd}, \overline{ac}, \overline{bd}$ are two-digit numbers. Given that

$$\overline{ab} + \overline{cd} = \overline{ac} + \overline{bd}.$$

What is the number of all possible quadruples (a, b, c, d)?

(A) 810 (B) 729 (C) 700 (D) 500 (E) 100

Problem 2.287. Let $A(a, b)$ and $B(a - b, a)$ be two points on the coordinate plane, where a and b are positive integers and $a > b$. Let O be the point $(0, 0)$ and the area of triangle ABO be 96. What is the value of $a + b$?

(A) 28 (B) 27 (C) 26 (D) 25 (E) 24

Problem 2.288. Let two cars start moving simultaneously toward each other from places A and B. Given that cars move with constant speeds and they cross each other in 2 hours. Given also that if these two cars start moving simulatenously toward each other from places A and B, such that each car moves with a constant speed that is greater by 10 km per hour than its initial speed, then they cross each other in 1 hour 48 minutes. What is the value of the distance (in kilometres) between A and B?

(A) 300 (B) 360 (C) 400 (D) 410 (E) 420

Problem 2.289. Let two circles of radii 6 be tangent to each other, such that each of these circles is tangent to three sides of trapezoid $ABCD$ with bases AD and BC. Given that $AB = 13$ and $CD = 20$. What is the value of the area of trapezoid $ABCD$?

(A) 198 (B) 396 (C) 300 (D) 342 (E) 210

Problem 2.290. Let ABC be a right triangle, such that $\angle B = 15°$ and the sum of the lenghts of legs AC, BC is equal to 10. Let CH be an altitude in triangle ABC and HE, HF be altitudes in triangles ACH, BCH, respectively. What is the value of the perimeter of rectangle $HECF$?

(A) 10 (B) 8 (C) 7 (D) 6 (E) 5

Problem 2.291. At most how many chess knights is possible to put on 4×4 grid square, such that each chess knight keeps under attack not more than two chess knights?

(A) 10 (B) 11 (C) 12 (D) 13 (E) 14

Problem 2.292. Given a square $ABCD$. Let M be a randomly chosen point in the interior part of square $ABCD$. What is the probability of the event that $\angle BMD \geq 135°$?

(A) $\dfrac{\pi - 2}{2}$ (B) $\dfrac{1}{2}$ (C) $\dfrac{\pi}{8}$ (D) $\dfrac{1}{4}$ (E) $\dfrac{\pi}{15}$

Problem 2.293. *Let figure Φ consists of a cylinder and a cone with the same base (see the figure). Given that the radius of the base is 5, the altitude of the cylinder is 10 and the altitude of the cone is 12. What is the value of the total surface area of figure Φ?*

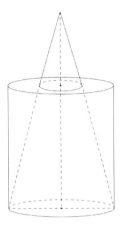

(A) 140π (B) 150π (C) $150\frac{1}{9}\pi$ (D) $151\frac{1}{9}\pi$ (E) 160π

Problem 2.294. *Let the sum of four pairwise different integers be equal to 2022. Given that the positive difference of each two of these numbers is prime number. What is the value of the greatest number among these four integers?*

(A) 409   (B) 507   (C) 508   (D) 509   (E) 510

Problem 2.295. *Let Φ be a figure on two dimensional coordinate plane consisting of points (x, y), such that*
$$|x^2 + y^2 - 1| + |(x - 1)^2 + y^2 - 1| \leq 3.$$
What is the value of the area of figure Φ?

(A) π (B) 2π (C) 2.25π (D) 10   (E) $10 + \dfrac{\pi}{4}$

Problem 2.296. *Let in a 2013×2013 square grid, some of the vertices of some of the cells be marked in red. Given that each of the cells of the square grid has at least two red vertices. What is the smallest possible number of red vertices?*

(A) 1006^2 (B) $2 \cdot 1007^2$ (C) 2000^2 (D) 1000   (E) 2000

Problem 2.297. *An ant moves from the bottom left corner to the top right corner of 6×8 rectangular grid. It can move only on the sides of unit cells, such that one move is either going up 1 unit or going down 1 unit or going right 1 unit. Given that the ant cannot pass twice the same side of any unit cell. What is the probability of the event that the ant passes through the center of symmetry of given 6×8 rectangular grid?*

(A) $\dfrac{18}{49}$ (B) $\dfrac{3}{7}$ (C) $\dfrac{31}{49}$ (D) $\dfrac{5}{7}$ (E) $\dfrac{6}{7}$

Problem 2.298. *What is the value of the sum of the smallest and the greatest solutions of the following equation belonging to $\left[0, \dfrac{3\pi}{2}\right]$?*
$$\cos x - \sin x + \cos 4x = -0.5.$$

(A) $\dfrac{2\pi}{3}$ (B) π (C) $\dfrac{4\pi}{3}$ (D) $\dfrac{3\pi}{2}$ (E) $\dfrac{5\pi}{3}$

Problem 2.299. *Let n be a positive integer and sequence x_n be defined as follows:*

$$x_1 = 1, x_2 = 2, x_3 = 5,$$

and

$$x_{n+3} = \frac{x_{n+2}^2 - x_{n+2}x_{n+1}}{x_{n+1} - x_n},$$

where $n = 1, 2, 3, \ldots$ What is the total number of all possible values of n for which $\frac{x_{n+1}}{x_n}$ is an integer?

(A) 1   (B) 2   (C) 4   (D) 2016   (E) 2017

Problem 2.300. *Let a, b, c be integers, such that $a \neq 0$ and*

$$3.9|a| \geq 2|b| + |c|.$$

Given that equation $ax^2 + bx + c = 0$ has only integer solutions. What is the total number of equations of the following form?

$$x^2 + \frac{b}{a}x + \frac{c}{a} = 0.$$

(A) 1   (B) 2   (C) 3   (D) 5   (E) 6

2.13 AMC 12 type practice test 13

Problem 2.301. What is the value of the expression $\dfrac{2^{2020} + 2^{2021}}{4^{1011} - 4^{1010}}$?

(A) $\dfrac{1}{2}$ (B) 1 (C) 2 (D) 4 (E) 2^{16}

Problem 2.302. Let n be a positive integer, such that $n! = 240 \cdot (n-2)!$. What is the value of n?

(A) 13 (B) 14 (C) 15 (D) 16 (E) 17

Problem 2.303. Given that the median of numbers 1, 2, x, 13 is equal to 3. What is the value of the mean of these numbers?

(A) 5 (B) 6 (C) 10 (D) 11 (E) 12

Problem 2.304. Let for positive real numbers x and y the operation $\star$ be defined in the following way $x \star y = \dfrac{1}{x} + \dfrac{1}{y}$. What is the value of the expression

$$\sqrt{1} \star \sqrt{2} - \sqrt{2} \star \sqrt{3} + \sqrt{3} \star \sqrt{4} - ... - \sqrt{98} \star \sqrt{99} + \sqrt{99} \star \sqrt{100}?$$

(A) 0 (B) 0.1 (C) 1 (D) 1.01 (E) 1.1

Problem 2.305. Given that 23 students are sitting in three rows, such that in each row there is at least one student. Given also that 20% of students of the first row, 25% of the second row and 10% of the third row are attending a basketball club. How many students from that class are attending a basketball club?

(A) 3 (B) 4 (C) 5 (D) 6 (E) 7

Problem 2.306. Given that the sum of three real numbers is 2 and their product is -2. Which of the following statements holds true?

(A) All three numbers are positive.
(B) All three numbers are negative.
(C) One of the numbers is 0.
(D) Two of the numbers are negative and the third is positive.
(E) Two of the numbers are positive and the third is negative.

Problem 2.307. What is the greatest number of all three-digit numbers with equal sums of the digits?

(A) 62 (B) 69 (C) 70 (D) 80 (E) 120

Problem 2.308. Let BE be a median of triangle ABC and CF be a median of triangle BEC. Given that $AF = AE$ and $CF = 20$. What is the length of line segment AB?

(A) 16 (B) 17 (C) 18 (D) 19 (E) 20

Problem 2.309. The factory did not work on any Saturdays or Sundays of February. On the nth day of each week it produced $(6-n)^2$ devices, where $n \in \{1, 2, 3, 4, 5\}$. Given that in total the factory produced 236 devices during the month of February. Which day of the week was the last day of that February?

(A) Monday (B) Tuesday (C) Wednesday (D) Thursday (E) Friday

Problem 2.310. *60 scientists took part in a seminar. Given that 50 of them speak English and 45 of them speak French. Given also that each participant speaks at least one of these two languages. What is the probability that randomly chosen two scientists can communicate with each other either in English or in French?*

(A) $\dfrac{54}{59}$ (B) $\dfrac{5}{59}$ (C) $\dfrac{1}{2}$ (D) $\dfrac{2}{3}$ (E) $\dfrac{3}{5}$

Problem 2.311. *Let point M be inside the rhombus $ABCD$. Given that $MA = AB = \sqrt{2+\sqrt{2}}$, $\angle MAB = 15°$ and $\angle ADC = 105°$. What is the length of line segment MC?*

(A) 1 (B) $\sqrt{2}$ (C) $\sqrt{3}$ (D) 2 (E) $\sqrt{2+\sqrt{5}}$

Problem 2.312. *Positive integer n is called "amazing", if the sum of its two greatest divisors is equal to 42. What is the total number of all "amazing" numbers?*

(A) 4 (B) 3 (C) 5 (D) 2 (E) 6

Problem 2.313. *What is the equation of the line that is symmetric to the line $y = 2x - 1$ with respect to line $y = x$?*

(A) $y = 2x + 1$ (B) $y = x - 0.5$ (C) $y = x + 0.5$ (D) $y = 0.5x + 0.5$ (E) $y = -x$

Problem 2.314. *Let $ABCD$ be a trapezoid with bases BC and AD, such that $BC = 20$ and $AD = 36$. Given that $AB = 63$ and $\angle ABC = 90°$. Let M be such a on leg AB that $BM = 15$. What is the value of the sum of the inradii of triangles ADM, BMC, CDM?*

(A) 25 (B) 27 (C) 29 (D) 30 (E) 32

Problem 2.315. *How many five-digit numbers ending with 12 are divisible by 13?*

(A) 23 (B) 27 (C) 60 (D) 68 (E) 69

Problem 2.316. *Let M be a given point in triangle ABC, such that $\angle MCA = 60°$, $AB = 3MC$, $AC : MB = 2 : \sqrt{3}$. Given that $\angle BAC = 60°$. What is the measure (in degrees) of angle MBA?*

(A) 5 (B) 10 (C) 15 (D) 30 (E) 45

Problem 2.317. *What is the greatest number of pairwise distinct positive integers such that the positive difference of each two of them is a prime number?*

(A) 3 (B) 4 (C) 5 (D) 6 (E) 10

Problem 2.318. *Consider all real values of a such that the equation $4^{x^2-5x} - 2^{x^2-5x} + a = 0$ has real solutions. For each of these values of a let $S(a)$ be sum of all real solutions of given equation. What is the greatest possible value of $S(a)$?*

(A) 2 (B) 4 (C) 5 (D) 10 (E) 25

Problem 2.319. Let $ABCD$ be a square with a side length of $a = 2\sqrt{1+\sqrt{3}}$. Consider the arcs of the circles with centers A, B, C, D and with radius $\dfrac{a}{2}$ which do not contain points outside the square (see the figure). Each of those 4 arcs are divided into 3 equal parts with 2 points. What is the area of the octagon with vertices at the above mentioned eight points?

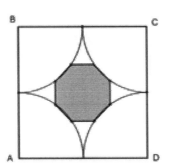

(A) $\dfrac{\sqrt{3}}{4}$ (B) $\dfrac{\sqrt{3}}{2}$ (C) 1 (D) 2 (E) 3

Problem 2.320. A three-digit number $\overline{abc}$ written using some of the digits 0, 1, 2, 3, 4 is randomly chosen. What is the probability that the inequalities $|a-b| \geq 2$ and $|b-c| \geq 2$ simultaneously hold true?

(A) 0.05 (B) 0.1 (C) 0.2 (D) 0.23 (E) 0.25

Problem 2.321. Let C_1 and B_1 be tangent points of the inscribed circle of triangle ABC with the sides AB and AC, respectively. Given that $\angle ACB = 92°$. Points M and N lie on the side AB, such that $C_1M = C_1N = CB_1$. What is the measure (in degrees) of angle MCN?

(A) 46 (B) 48 (C) 45 (D) 50 (E) 44

Problem 2.322. What is the sum of all positive integers n less than 100, such that for each of them the equality $\dfrac{1}{n} + \dfrac{1}{n+2} = 0.\overline{a_n b_n c_n d_n e_n f_n}$ holds true, where $a_n, b_n, c_n, d_n, e_n, f_n$ are some digits?

(A) 61 (B) 69 (C) 63 (D) 65 (E) 67

Problem 2.323. What is the sum of the greatest and smallest solutions of the equation $\cos x - \sin x + \cos 4x = -0.5$ in the interval $\left[0, \dfrac{5\pi}{3}\right]$?

(A) $\dfrac{2\pi}{3}$ (B) π (C) $\dfrac{4\pi}{3}$ (D) $\dfrac{3\pi}{2}$ (E) $\dfrac{5\pi}{3}$

Problem 2.324. Let $\lambda(n)$ be the number of all divisors of n. Given that first $f(n)$ factors form an arithmetic sequence, but the first $f(n)+1$ factors do not. What is the greatest possible value of $2f(n) - \lambda(n)$?

(A) -1 (B) 0 (C) 1 (D) 2 (E) 7

Problem 2.325. Let set $S = \{(i,j) \mid i \in \{1,2,...,10\}$ and $j \in \{1,2,...,10\}\}$ is given on the coordinate plane. Triangle ABC is chosen on the plane, such that none of the points A, B, C belong to S. Let n be the number of points in set S, such that each of them lies on one of the sides of triangle ABC. What is the greatest possible value of n?

(A) 10 (B) 20 (C) 22 (D) 23 (E) 30

2.14 AMC 12 type practice test 14

Problem 2.326. What is the value of the expression $(15^{-2} + 20^{-2})^{-1} - (15^{-1} - 20^{-1})^{-1}$?

(A) 144 (B) 100 (C) 84 (D) 48 (E) 4

Problem 2.327. David and Anna ate 7 candies altogether. After David ate another 5 candies, the number of candies that he ate was twice the number of candies that Anna ate. How many candies did Anna eat?

(A) 3 (B) 4 (C) 5 (D) 6 (E) 2

Problem 2.328. What is the value of the sum $0.10 + 0.11 + ... + 0.99$?

(A) 49 (B) 49.01 (C) 49.03 (D) 49.05 (E) 49.5

Problem 2.329. What is the sum of all positive integers n such that $n!$ ends with exactly 6 zeros?

(A) 135 (B) 107 (C) 78 (D) 51 (E) 25

Problem 2.330. A two-digit number is 8 times greater than the sum of its digits. What is the sum of all digits of this two-digit number?

(A) 5 (B) 7 (C) 9 (D) 11 (E) 12

Problem 2.331. Given that the sum of the ages of all students in the classroom is 120. After 6 years the sum of their ages will be twice as much as the sum of their ages 3 years ago. What is the value of the arithmetic mean of the ages of all students?

(A) 8 (B) 10 (C) 12 (D) 15 (E) 16

Problem 2.332. In a geometric sequence with positive terms, the difference between the seventh and the first terms is seven times more than the sum of the fourth and the first terms. What is the common ratio of this geometric sequence?

(A) 1 (B) $\sqrt[3]{2}$ (C) 2 (D) 3 (E) 3.2

Problem 2.333. David solves five problems each Saturday and Sunday, and he solves six problems each weekday. During some consecutive days, David solved 70 problems. Which day of the week did he start solving the problems?

(A) Wednesday (B) Thursday (C) Friday (D) Sunday (E) Monday

Problem 2.334. Let $ABCD$ be a square with a side length $\sqrt{3}$. Let square $ABCD$ be rotated around point A by $30°$ and as a result we obtained square $AB'C'D'$. What is the area of the part that squares $AB'C'D'$ and $ABCD$ have in common?

(A) 1 (B) 1.5 (C) $\sqrt{3}$ (D) 2 (E) 2.1

Problem 2.335. Let us choose any random moment between 12 AM and 12 PM. What is the probability of the event that all three arrows of the clock are located on the right semi-circle of the clock?

(A) $\frac{1}{2}$ (B) $\frac{1}{3}$ (C) $\frac{1}{8}$ (D) $\frac{1}{12}$ (E) $\frac{1}{15}$

Problem 2.336. Given that $2\sin x + \cos x = \sqrt{5}$. What is the value of $\tan x$?

(A) $\sqrt{2}$ (B) 2 (C) $\sqrt{3}$ (D) 1 (E) $\frac{\sqrt{5}}{5}$

Problem 2.337. *Three sisters bought 4 identical bracelets. In how many different ways can they wear those 4 bracelets? (they can wear each bracelet either on right or left hand).*

(A) 15   (B) 20   (C) 24   (D) 120   (E) 126

Problem 2.338. *Let ABC be a triangle, such that $AB = 12$, $BC = 9$, $AC = 16$. Given that point D lies on the side AC and point E lies on the ray DB, so that $CD = 7$ and $\angle DAE = \angle ABD$. What is the length of line segment AE?*

(A) 9   (B) 11   (C) 12   (D) 16   (E) 18

Problem 2.339. *Let p be a prime number, such that $4p - 113$ and $4p + 113$ are prime numbers. What is the value of the sum of all the digits of p?*

(A) 10   (B) 11   (C) 21   (D) 25   (E) 30

Problem 2.340. *Let $ABCD$ be a square with a side lenght of $2\sqrt{5}$. Let M, N, P, K be the midpoints of sides AB, BC, CD, AD, respectively. What is the radius of the circle that is tangent to each of line segments AN, DN, AP, BP, BK, CK, CM, DM?*

(A) $\dfrac{\sqrt{5}}{2}$ (B) 1   (C) $\dfrac{\sqrt{3}}{2}$ (D) $\dfrac{\sqrt{2}}{2}$ (E) $\dfrac{1}{2}$

Problem 2.341. *At least 11 mathematicians took part in the Math conference. Given that each participant is acquainted with no more than 11 participants. Given also that for each group of 10 participants (of the conference) there exists a participant in the conference who is acquainted with all these 10 participants. What is the greatest number of the participants in this Math conference?*

(A) 11   (B) 12   (C) 13   (D) 14   (E) 15

Problem 2.342. *A circle with center I is inscribed in the pentagon $ABCDE$. Given that $\angle BCD = 100°$. Let $\angle AIB + \angle EID = n°$. What is the value of n?*

(A) 100   (B) 120   (C) 140   (D) 150   (E) 160

Problem 2.343. *Each side of a convex hexagon $ABCDEF$ is colored either white, blue or red. Given that each two sides with common vertex have different colors. In how many different ways is it possible to color all sides of this hexagon?*

(A) 36   (B) 48   (C) 66   (D) 97   (E) 98

Problem 2.344. *Let ABC be a triangle, such that $AB < BC$. Let D be the midpoint of the arc AC of the circumcircle of ABC, such that points B and D are on the different sides of line AC. Let line segment DE is perpendicular to the chord BC and point E belongs to chord BC. Given that $BE = 17$ and $EC = 7$. What is the length of side AB?*

(A) 7   (B) 8   (C) 9   (D) 10   (E) 11

Problem 2.345. *For which value of a the equation $\log_a(x^2 - 4x + a + 4) = \log_{a^2 - 3a + 7}(8x - 2x^2 - 3)$ has only one solution?*

(A) 6   (B) 5   (C) 4   (D) 3   (E) 2

Problem 2.346. *What is the sum of all three-digit numbers, such that for each of them the sum of its digits is divisible by 7?*

(A) 2000   (B) 3500   (C) 50000   (D) 69237   (E) 75730

Problem 2.347. Let AD be an angle bisector of triangle ABC. Given that $BD = 7$, $CD = 8$, $AC - AB = 13$ and $\angle A + 2\angle C = n°$. What is the value of n?

(A) 60 (B) 75 (C) 90 (D) 105 (E) 120

Problem 2.348. What is the sum of all digits of the greatest integer, which is not possible to express as a sum of two positive integers, such that each of them is divisible by a square of an integer greater than 1?

(A) 2 (B) 4 (C) 5 (D) 7 (E) 11

Problem 2.349. Let M be the greatest possible value of the expression $|\sin x(1-\cos y)| + |\sin y(1-\sin x)|$. What is the value of $(M-1)^2$?

(A) 4 (B) 5 (C) 6 (D) 6.1 (E) 6.2

Problem 2.350. Let $p(x)$, $q(x)$, $r(x)$ be polynomials of degree 4 with real coefficients, such that for each real value of x it holds true $\sqrt{p(x)} + \sqrt{q(x)} = \sqrt{r(x)}$. Given that $p(1) = 0$, $q(2) = 0$, $r(1) = 4$, $r(2) = 1$. What is the value of $r(4)$?

(A) 1 (B) 16 (C) 100 (D) 289 (E) 400

Chapter 3

Answers

3.1 Answers of AMC 12 type practice tests

Problem	Practice test 1	Practice test 2	Practice test 3	Practice test 4
1	B	B	D	E
2	A	A	C	A
3	E	C	E	C
4	E	B	D	B
5	B	E	C	D
6	E	D	D	B
7	B	E	A	E
8	D	B	B	C
9	B	A	A	E
10	C	C	D	B
11	C	D	E	C
12	E	B	C	A
13	A	D	D	D
14	B	E	B	D
15	E	C	A	A
16	E	E	C	B
17	B	E	D	D
18	C	A	C	C
19	A	C	E	D
20	A	D	B	E
21	D	B	A	A
22	D	E	C	C
23	C	D	B	B
24	E	C	D	E
25	A	A	E	C

Problem	Practice test 5	Practice test 6	Practice test 7	Practice test 8
1	D	D	B	C
2	D	C	A	A
3	B	C	E	B
4	B	E	A	E
5	B	D	A	B
6	E	B	B	B
7	A	A	A	C
8	C	B	E	B
9	C	B	A	D
10	E	D	A	D
11	A	C	A	B
12	B	E	C	E
13	C	C	A	E
14	E	A	A	C
15	D	E	B	A
16	A	B	E	D
17	E	E	B	A
18	C	C	C	C
19	B	A	C	B
20	B	C	B	C
21	D	B	A	B
22	A	C	E	C
23	C	E	D	B
24	E	D	C	A
25	B	B	B	E

Problem	Practice test 9	Practice test 10	Practice test 11	Practice test 12
1	C	A	B	D
2	A	C	D	B
3	C	B	B	C
4	E	C	A	B
5	E	E	D	D
6	D	D	A	D
7	A	A	E	D
8	D	B	D	B
9	C	C	E	C
10	D	C	C	A
11	E	E	A	A
12	C	D	C	E
13	A	D	C	B
14	C	A	E	D
15	E	B	B	E
16	B	C	B	C
17	B	E	C	A
18	A	A	C	D
19	C	D	E	D
20	B	B	C	C
21	A	B	B	B
22	C	A	B	C
23	E	B	C	D
24	B	E	B	B
25	C	C	D	E

Problem	Practice test 13	Practice test 14
1	B	C
2	D	B
3	A	D
4	E	A
5	B	C
6	E	C
7	C	C
8	E	E
9	B	C
10	A	C
11	B	B
12	B	E
13	D	D
14	B	B
15	E	B
16	D	B
17	B	C
18	D	C
19	D	D
20	D	E
21	E	D
22	D	A
23	D	C
24	D	B
25	B	D

Chapter 4

Solutions

4.1 Solutions of AMC 12 type practice test 1

Problem 4.1. *Alice needs to solve exactly 100 problems in n days. Given that each day she can solve either 5, 6 or 7 problems. What is the smallest possible value of n?*

(A) 14   (B) 15   (C) 20   (D) 16   (E) 17

Solution. Answer. (B)
Note that even if Alice solves 7 problems per day, she will need more than 14 days to solve at least 100 problems, as $\frac{100}{7} > 14$. Thus, it follows that $n \geq 15$.
On the other hand, for $n = 15$ we can provide the following example $12 \cdot 7 + 1 \cdot 6 + 2 \cdot 5 = 100$.
Therefore, the smallest possible value of n is equal to 15. □

Problem 4.2. *The sum of the squares of two positive numbers is three times their product. How many times is the square of the sum of these numbers greater than the square of their difference?*

(A) 5   (B) 3   (C) 9   (D) 2   (E) 4

Solution. Answer. (A)
Let us suppose that those positive numbers are a and b. Given that
$$a^2 + b^2 = 3ab.$$
Thus, it follows that
$$\frac{(a+b)^2}{(a-b)^2} = \frac{a^2 + b^2 + 2ab}{a^2 + b^2 - 2ab} = \frac{5ab}{ab} = 5.$$
Therefore, the square of the sum of these numbers is 5 times greater than the square of their difference. □

Problem 4.3. *A student was assigned a test consisting of 10 problems. Each correct answer was scored 2 points, an incomplete answer was scored 1 point, and no point was scored for a wrong answer. In the end of the test the total score of the student was 19. Which of the following statements is true?*

(A) The student correctly answered to all 10 problems.
(B) At least one of the answers was incorrect.
(C) All the answers for all 10 problems were incomplete.
(D) Only one answer was correct.
(E) Only one answer was incomplete.

Solution. Answer. (E)
(A) is not true, as the total score would have been $10 \cdot 2 = 20$ points.
(B) is not true, as the total score could not be more than $9 \cdot 2 = 18$ points.
(C) is not true, as the total score would have been $10 \cdot 1 = 10$ points.
(D) is not true, as the total score could not be more than $9 \cdot 1 + 2 = 11$ points.
(E) is true, as $9 \cdot 2 + 1 = 19$. □

Problem 4.4. *By what percent is the circumference of the incircle of a square less than the perimeter of that square?*

(A) 50 (B) 22 (C) 21 (D) $\dfrac{100(4-\pi)}{\pi}$ (E) $25(4-\pi)$

Solution. Answer. (E)
Let the side length of the square be a, then the perimeter of the square is equal to $4a$.
Note that the inradius of the square is equal to $\dfrac{a}{2}$ and the circumference of the incircle is equal to πa.
Therefore, the circumference of the incircle of the square is less than the perimeter of the square by the following percent:
$$\frac{4a - \pi a}{4a} \cdot 100 = 25(4-\pi).$$
□

Problem 4.5. *Participants of a mathematics conference stay in two hotels. Participants staying in the same hotel shook hands with each other exactly once, while participants staying in different hotels did not. The total number of handshakes is the product of the number of participants in each hotel. Given that the total number of participants is greater than 17 and less than 34. What is the total number of participants?*

(A) 20 (B) 25 (C) 18 (D) 30 (E) 33

Solution. Answer. (B)
Let m and n be the number of participants staying in these hotels, then the total number of handshakes is equal to:
$$\binom{m}{2} + \binom{n}{2} = \frac{m(m-1)}{2} + \frac{n(n-1)}{2}.$$
Given that
$$\frac{m(m-1)}{2} + \frac{n(n-1)}{2} = mn.$$
Therefore $m + n = (m-n)^2$ and $17 < m+n < 34$.
We obtain that $m + n = 25$.
Hence, the total number of participants is 25. □

Problem 4.6. *Let $ABCD$ be a square of side length 12. Let M be an interior point of $ABCD$. Given that the distance from M to side AB, AD, CD is a, b, c, respectively. How many possible points M are there, such that a, b, c are integers and there exists a triangle with sides a, b, c?*

(A) 72 (B) 60 (C) 66 (D) 59 (E) 61

Solution. Answer. (E)
All such points are shown in the figure below and their total number is $2(1+3+5+7+9+11) - 11 = 61$.

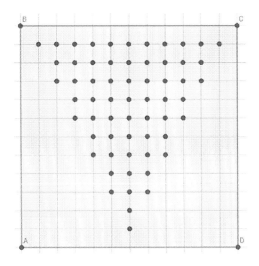

Therefore, there are 61 possible such points M. □

Problem 4.7. *Let n be a positive integer and function $f(n)$ be defined as follows:*

$$f(1) = 5,$$

$$f(n) = f\left(\frac{n-1}{2}\right) + 1, if\ n\ is\ odd\ greater\ than\ 1,$$

$$f(n) = f\left(\frac{n}{2}\right), if\ n\ is\ even.$$

What is the value of $f(2017)$?

(A) 9   (B) 11   (C) 10   (D) 1008   (E) 20

Solution. Answer. (B)
Given that

$$f(2017) = f(1008) + 1 = f(504) + 1 = f(252) + 1 = f(126) + 1 = f(63) + 1 = f(31) + 2 =$$
$$= f(15) + 3 = f(7) + 4 = f(3) + 5 = f(1) + 6 = 11.$$

Thus, it follows that $f(2017) = 11$.
Alternative solution.
By using mathematical induction we can prove that

$$n = 2^{k_1} + 2^{k_2} + ... + 2^{k_s},$$

where $k_1 > k_2 > ... > k_s \geq 0, k_1, k_2, ..., k_s$ are integers.
Thus, it follows that $f(n) = s + 4$. Given that

$$2017 = 2^{10} + 2^9 + 2^8 + 2^7 + 2^6 + 2^5 + 1.$$

Therefore $f(2017) = 7 + 4 = 11$. □

Problem 4.8. *Given a cube with an edge length of 1 unit. Let Φ be the solid formed by all points that are located at a distance not greater than 1 unit from some point on the surface of the cube. What is the volume of Φ?*

(A) 27   (B) $7 + 3\pi$ (C) $7 + \dfrac{4\pi}{3}$ (D) $7 + \dfrac{13\pi}{3}$ (E) $6 + 4\pi$

Solution. Answer. (D)
The solid Φ consists of 7 unit cubes, 12 solids each 4 of which create a cylinder of radius 1 and height 1, and also 8 solids which create a sphere of radius 1.

Therefore, the volume of solid Φ is $7 + 3\pi + \dfrac{4\pi}{3} = 7 + \dfrac{13\pi}{3}$. $\square$

Problem 4.9. *What is the locus of all possible points (x, y) satisfying the following inequality?*
$$|x+y-1| + |x-y+1| + |x+y+1| + |x-y-1| \le 4$$

(A) four vertices of a square.
(B) a square and its inner region.
(C) a triangle and its inner region.
(D) three vertices of a triangle.
(E) eight points.

Solution. Answer. (B)
Let that image be φ. It is known that $|a| + |b| \ge |a+b|$. Moreover $|a| + |b| = |a+b|$, when $ab \ge 0$. According to this inequality, we have that

$$4 \ge |x+y-1| + |x-y-1| + |x+y+1| + |x-y+1| \ge |2x-2| + |2x+2| = |2-2x| + |2x+2| \ge 4.$$

Thus, it follows that
$$|2-2x| + |2+2x| = 4,$$
$$|1-x-y| + |1-x+y| = |2-2x|,$$
$$|1+x+y| + |1+x-y| = |2+2x|.$$

We obtain that φ is the locus of all possible points (x, y) for which
$$\begin{cases} 2 - 2x \ge 0, \\ 2 + 2x \ge 0, \\ 1 - x - y \ge 0, \\ 1 - x + y \ge 0, \\ 1 + x + y \ge 0, \\ 1 + x - y \ge 0. \end{cases}$$

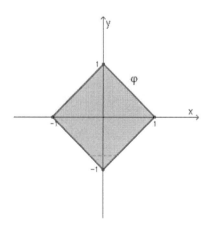

Therefore, the locus of all possible points (x, y) satisfying given inequality is a square and its inner region. $\square$

Problem 4.10. *Let the distance between the centers of two circles of radii 8 and 2 be equal to 3. A circle of radius 1 is randomly placed within the circle of radius 8. What is the probability that the circles of radii 1 and 2 intersect?*

(A) $\dfrac{1}{16}$ (B) $\dfrac{9}{64}$ (C) $\dfrac{9}{49}$ (D) $\dfrac{1}{5}$ (E) $\dfrac{1}{2}$

Solution. Answer. (C)
Note that in order the circle of radius 1 to be in the circle of radius 8, its center should lie in the circle of radius 7 (see the figure).

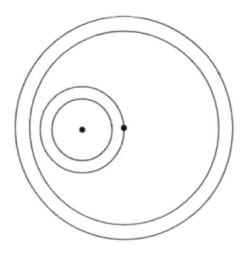

Moreover, in order the circles of radii 1 and 2 to have a common point, the center of the circle of radius 1 should lie in the circle of radius 3.
Therefore, the probability that the circles of radii 1 and 2 intersect is equal to $\dfrac{9}{49}$. □

Problem 4.11. *Let some internal angles of a convex polygon be acute. Given that the sum of its non-acute angles is equal to 2017° and the sum of its acute angles is equal to n°. What is the value of n?*

(A) 323 (B) 117 (C) 143 (D) 163 (E) 37

Solution. Answer. (C)
Let m be the number of sides of this convex polygon. Hence, the sum of its internal angles is equal to $180°(m-2)$.
On the other hand, the sum of its external angles is equal to $360°$.
Therefore, the number of its acute angles is not greater than 3.
Therefore, $n < 270$ and $180 \mid 2017 + n$.
Thus, it follows that $n = 143$. □

Problem 4.12. *A positive number is called "nice", if at least six of its divisors are from the set*

$$\{1, 2, 3, 4, 5, 6, 7, 8, 9, 10\}.$$

What is the smallest possible value of the positive difference of two "nice" numbers?

(A) 1 (B) 3 (C) 6 (D) 4 (E) 2

Solution. Answer. (E)
Let number a and b be "nice" numbers and $a > b$. Note that both a and b have at least two common divisors. Thus, it follows that there exists a positive integer c greater than 1, such that $c \mid a$ and $c \mid b$. Therefore $c \mid (a - b)$.
Hence, we obtain that $a - b \geq c \geq 2$. Thus, it follows that $a - b \geq 2$.
Note that 882 and 880 are "nice" numbers and their positive difference is 2. $\square$

Problem 4.13. *Sophia planned to cover a distance of 160 miles at a constant rate. Given that half of the distance she moved 5 miles per hour faster than she has planned, and for the rest of the distance she moved 4 miles per hour slower than she has planned. It turned out that Sophia has spent as much time covering the total distance as she has planned. In how much time (in hours) Sophia has planned to cover the total distance?*

(A) 4   (B) 8   (C) 3   (D) 2   (E) 5

Solution. Answer. (A)
Let us denote Sophia's planned rate to cover the total distance of 160 miles by x miles per hour.
Thus, it follows that she has planned to cover the total distance in $\dfrac{160}{x}$ hours. Given that half of the distance she moved 5 miles per hour faster than she has planned, and for the rest of the distance she moved 4 miles per hour slower than she has planned. Hence, we deduce that

$$\frac{160}{x} = \frac{80}{x+5} + \frac{80}{x-4}.$$

Thus, it follows that
$$2(x+5)(x-4) = x(2x+1).$$

Hence $x = 40$ and $\dfrac{160}{40} = 4$.
Therefore, Sophia has planned to cover the total distance in 4 hours. $\square$

Problem 4.14. *Let five chairs be placed around a circular table. In how many different ways can two girls and three boys be seated, such that two girls do not sit next to each other?*

(A) 120   (B) 60   (C) 30   (D) 20   (E) 24

Solution. Answer. (B)
The first girl can be seated in 5 ways. Given that the second girl cannot sit next to the first girl, therefore the second girl can be seated in 2 ways.
Note that once the girls take their seats the boys can take their seats in 3! ways.
Therefore, the answer is $5 \cdot 2 \cdot 3! = 60$. $\square$

Problem 4.15. *What is the smallest integer value of n greater than 1, such that the following inequality holds true?*
$$\sin n + 2\cos n + \tan n > 0.$$

(A) 2   (B) 3   (C) 4   (D) 5   (E) 6

Solution. Answer. (E)
Note that
$$\sin 2 + 2\cos 2 + \tan 2 = \frac{\sin 2(1 + \cos 2) + 2\cos^2 2}{\cos 2} < 0,$$

as $\sin 2 > 0, 1 + \cos 2 > 0$ and $\cos 2 < 0$.
In a similar way, we obtain that $\sin 3 + 2\cos 3 + \tan 3 < 0$.

We have that $0 < 4 - \pi < \dfrac{\pi}{3}$, therefore

$$\sin 4 + 2\cos 4 + \tan 4 = -\sin(4 - \pi) - 2\cos(4 - \pi) + \tan(4 - \pi) <$$
$$< -\sin(4 - \pi) - 2\cos(4 - \pi) + 2\sin(4 - \pi) = \sin(4 - \pi) - 2\cos(4 - \pi) < \sin(4 - \pi) - 1 < 0.$$

Thus, it follows that $\sin 4 + 2\cos 4 + \tan 4 < 0$.

Note that
$$\dfrac{3\pi}{2} < 5 < \dfrac{3\pi}{2} + \dfrac{\pi}{6}.$$

Therefore $\cos 5 < \dfrac{1}{2}$ and $\tan 5 < -\sqrt{3}$. Thus, it follows that

$$\sin 5 + 2\cos 5 + \tan 5 < 0 + 1 - \sqrt{3} < 0.$$

Note that
$$2\pi - \dfrac{\pi}{6} < 6 < 2\pi.$$

We deduce that
$$\sin 6 > -\dfrac{1}{2}, \cos 6 > \dfrac{\sqrt{3}}{2}, \tan 6 > -\dfrac{1}{\sqrt{3}}.$$

Hence, we obtain that
$$\sin 6 + 2\cos 6 + \tan 6 > -\dfrac{1}{2} + \sqrt{3} - \dfrac{\sqrt{3}}{3} > 0.$$

Therefore, the smallest integer value of n greater than 1, such that given inequality holds true is equal to 6. $\square$

Problem 4.16. *Let circles of radii 1, 2, 3 be pairwise externally tangent. What is the radius of the circle that is internally tangent to each of the three circles?*

(A) $2\dfrac{5}{23}$ (B) $4\dfrac{10}{23}$ (C) $8\dfrac{20}{23}$ (D) 5   (E) 6

Solution. Answer. (E)

Let us denote the centers of the circles of radii 1, 2 and 3 by O, A, B, respectively. Then $OA = 3, OB = 4, AB = 5$.

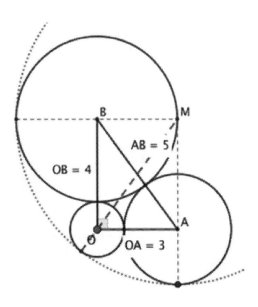

Note that
$$AB^2 = 5^2 = 3^2 + 4^2 = OA^2 + OB^2.$$
Thus, it follows that $\angle AOB = 90°$. Let R be the radius of the circle that is internally tangent to each of the three circles and let M be its center. Given that $OM = R - 1$, $AM = R - 2$ and $BM = R - 3$. Let us construct rectangular coordinate system, such that $O(0,0)$, $A(0,3)$, $B(4,0)$ and let $M(x, y)$. Then, we have that
$$x^2 + y^2 = (R - 1)^2.$$
$$x^2 + (y - 3)^2 = (R - 2)^2.$$
$$(x - 4)^2 + y^2 = (R - 3)^2.$$
From the first two equations, we deduce that $y = \dfrac{R}{3} + 1$.

On the other hand, from the first and the last equations we deduce that $x = \dfrac{R}{2} + 1$.

Thus, it follows that
$$\left(\frac{R}{3} + 1\right)^2 + \left(\frac{R}{2} + 1\right)^2 = (R - 1)^2.$$
Therefore $R = 6$. □

Problem 4.17. *What is the total number of all complex numbers z satisfying the following two conditions?*
$$z^{12} = 1,$$
$$(z - \sqrt{3})^{36} = 1.$$

(A) 0   (B) 2   (C) 6   (D) 4   (E) 3

Solution. Answer. (B)
From given equations we obtain that $|z| = 1$ and $|z - \sqrt{3}| = 1$, so complex number z can be the point of intersection of two circles: with center (0,0) and of radius 1, and with center $(\sqrt{3}, 0)$ and of radius 1. Therefore, their total number is not more than two.
On the other hand, each of the numbers
$$\cos\frac{\pi}{6} + i\sin\frac{\pi}{6},$$
and
$$\cos\frac{\pi}{6} - i\sin\frac{\pi}{6},$$
satisfies the condition of the problem, as
$$\left(\cos\frac{\pi}{6} \pm i\sin\frac{\pi}{6}\right)^{12} = \cos 2\pi \pm i\sin 2\pi = 1,$$
and
$$\left(\cos\frac{\pi}{6} \pm i\sin\frac{\pi}{6} - \sqrt{3}\right)^{36} = \left(-\cos\frac{\pi}{6} \pm i\sin\frac{\pi}{6}\right)^{36} = \cos 6\pi \pm i\sin 6\pi = 1.$$
Thus, it follows that the total number of all complex numbers z satisfying given two conditions is equal to 2. □

Problem 4.18. Let $n = \overline{a_1 a_2 ... a_k}$ and
$$T(n) = |a_1 - a_2 + ... + (-1)^{k-1} a_k|.$$
For example
$$T(1237) = |1 - 2 + 3 - 7| = 5.$$
Given that $T(n) = 4$ for some positive integer n. Which of the following values can be equal to $T(n-1)$?

(A) 2   (B) 9   (C) 6   (D) 1   (E) 7

Solution. Answer. (C)
Note that
$$T(n) = |a_1 - a_2 + ... + (-1)^{k-1} a_k| = |(-1)^{k-1} a_1 + (-1)^{k-2} a_2 + ... + a_k|,$$
and
$$\overline{a_1 a_2 ... a_k} - ((-1)^{k-1} a_1 + (-1)^{k-2} a_2 + ... + a_k) = a_1(10^{k-1} - (-1)^{k-1}) + a_2(10^{k-2} - (-1)^{k-2}) + ... + a_{k-1}(10 - (-1)).$$

Obviously
$$11 \mid 10^m - (-1)^m,$$
for any positive integer m. Thus, it follows that
$$11 \mid n - T(n),$$
or
$$11 \mid n + T(n).$$
If $T(n)$ leaves a remainder of 4 after division by 11, then $n = 11k + 4$ or $n = 11k + 7$, for some positive integer k. Therefore
$$n - 1 = 11k + 3,$$
or
$$n - 1 = 11k + 6.$$
Note that $T(n-1)$ can leave a remainder of 3, 8, 6 or 5 after division by 11. Hence, we obtain that $T(n-1)$ can leave only a remainder of 6 after division by 11.
Example. If $n = 3010$, then $T(n-1) = 6$. □

Problem 4.19. *Let a square with a side length of x be inscribed into a triangle with side lengths of 13, 14, 15, such that two of its vertices lie on the smallest side of the triangle. Let another square with a side length of y be inscribed into another triangle which is congruent to the first triangle, such that two of its vertices lie on the largest side of the triangle. What is the value of $\dfrac{15}{x} - \dfrac{13}{y}$?*

(A) $\dfrac{56}{195}$ (B) 1   (C) $\dfrac{1685}{1703}$ (D) $\dfrac{2}{3}$ (E) $\dfrac{3}{2}$

Solution. Answer. (A)
Let $MNPQ$ be the square inscribed in triangle ABC (see the figure).
Given that $AB = c, BC = a, AC = b$ and $MN = d$.
Let $BE \perp AC, BE = h_b$.

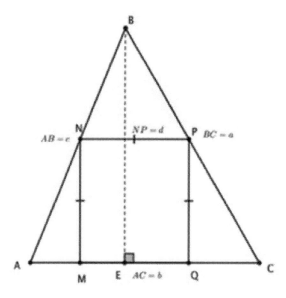

We have that $NP \parallel AC$. Thus, it follows that $\angle BNP = \angle BAC$. Hence, we obtain that $\triangle BNP \sim \triangle BAC$. Therefore
$$\frac{h_b - d}{h_b} = \frac{d}{b}.$$

We deduce that
$$d = \frac{2S}{b + h_b},$$
where S is the area of the triangle ABC. Thus, it follows that
$$d = \frac{2Sb}{b^2 + 2S}.$$

Taking this into consideration, we deduce that
$$\frac{15}{x} - \frac{13}{y} = 15\left(\frac{13}{2S} + \frac{1}{13}\right) - 13\left(\frac{15}{2S} + \frac{1}{15}\right) = \frac{15}{13} - \frac{13}{15} = \frac{56}{195}.$$

□

Problem 4.20. *For how many real numbers a the values of the following three expressions form a geometric sequence?*
$$\log_2 a, \log_2(a^2 - 1), -\log_2 \frac{(a^2 - 1)^{\sqrt{2}-1}}{a^{\sqrt{2}}}.$$

(A) 2 (B) 3 (C) 1 (D) 4 (E) 0

Solution. Answer. (A)
Given that
$$\left(\log_2(a^2 - 1)\right)^2 = -\log_2 a\left((\sqrt{2} - 1)\log_2(a^2 - 1) - \sqrt{2}\log_2 a\right),$$
or
$$\left(\log_2(a^2 - 1) - \log_2 a\right)\left(\log_2(a^2 - 1) + \sqrt{2}\log_2 a\right) = 0.$$

Thus, it follows that either
$$\log_2(a^2 - 1) = \log_2 a,$$
or
$$\log_2(a^2 - 1) = \log_2 \frac{1}{a^{\sqrt{2}}}.$$
Hence, we obtain that either $a^2 - a - 1 = 0$ or $a^2 - 1 = \frac{1}{a^{\sqrt{2}}}$, where $a > 1$.
Note that, by constructing the graphs of functions $y = x^2 - 1$ and $y = \frac{1}{x^{\sqrt{2}}}$ one can easily verify that equation $a^2 - 1 = \frac{1}{a^{\sqrt{2}}}$ has a unique solution.
On the other hand, note that $a = \frac{1+\sqrt{5}}{2}$ is a solution.
Therefore, for two real numbers a the values of given three expressions form a geometric sequence. $\square$

Problem 4.21. *Let a, b, c, d be integers, such that $ad \neq 0$ and 5 is a root of the polynomial $ax^3 + bx^2 + cx + d$. What is the smallest possible value of $|a| + |b| + |c| + |d|$?*

(A) 4   (B) 5   (C) 6   (D) 12   (E) 14

Solution. Answer. (D)
Given that
$$125a + 25b + 5c + d = 0.$$
Thus, it follows that $5 \mid d$ and
$$a + b + c + d = -124a - 24b - 4c.$$
Hence, we obtain that $a + b + c + d$ is even. Therefore $|a| + |b| + |c| + |d|$ is even.
We have that $d = 5m$, where m is a nonnegative integer. Given that
$$|a| + |b| + |c| + |d| = |a| + |b| + |-25a - 5b - m| + |5m| \geq$$
$$\geq |a| + |b| + |-5a - b - \frac{m}{5}| + 5|m| \geq |a| + |5a + \frac{m}{5}| + |5m| \geq 1 + 4\frac{4}{5} + 5 = 10\frac{4}{5}.$$
Thus, it follows that $|a| + |b| + |c| + |d| \geq 12$.
Example. Note that $x = 5$ is a root of the polynomial $x^3 - 5x^2 + x - 5$ and $|1| + |-5| + |1| + |-5| = 12$.
Therefore, the smallest possible value of $|a| + |b| + |c| + |d|$ is equal to 12. $\square$

Problem 4.22. *A particle moves with a constant speed on a unit square grid. Every second, it moves from point (x, y) either to point $(x + 1, y)$ or to point $(x, y + 1)$, where x, y are nonnegative integers. The particle starts at point (0, 0) and ends its route at point (4,4). What is the probability that the route passes through point (2,2)?*

(A) $\frac{1}{2}$ (B) $\frac{3}{8}$ (C) $\frac{17}{35}$ (D) $\frac{18}{35}$ (E) $\frac{5}{8}$

Solution. Answer. (D)
First let us find the number of all paths. Note that the particle moves either up or right, so every path may be replaced by a word consisting of eight letters in which four letters are "u" and the other four are "r". Therefore, the total number of paths is equal to $\binom{8}{4}$.
On the other hand, the total number of paths passing through point (2,2) is equal to $\binom{4}{2} \cdot \binom{4}{2}$.
Therefore, the probability that the route passes through point (2,2) is:
$$\frac{\binom{4}{2} \cdot \binom{4}{2}}{\binom{8}{4}} = \frac{18}{35}.$$

$\square$

Problem 4.23. *Given that there exists a rational number k, such that each of the polynomials*
$$x^3 + x^2 + kx + 2,$$
and
$$x^4 - 4x^3 + 4x^2 + (k+5)x - 3,$$
has an integer root larger than 1. What is the value of k?

(A) $\dfrac{5}{24}$ (B) $-\dfrac{3}{25}$ (C) -7   (D) 70   (E) 3

Solution. Answer. (C)
Let m and n be the roots (greater than 1) of those polynomials, in that case
$$k = -m^2 - m - \frac{2}{m}$$
and
$$k + 5 = -n^3 + 4n^2 - 4n + \frac{3}{n}.$$
Thus, it follows that
$$\frac{3}{n} + \frac{2}{m} = n^3 - 4n^2 + 4n - m^2 - m + 5.$$
Hence, we obtain that
$$\frac{3}{n} + \frac{2}{m}$$
is an integer.
On the other hand, we have that
$$\frac{3}{n} + \frac{2}{m} \le \frac{3}{2} + 1 = 2.5.$$
Hence, we obtain that either
$$\frac{3}{n} + \frac{2}{m} = 1,$$
or
$$\frac{3}{n} + \frac{2}{m} = 2.$$
We deduce that either
$$(n-3)(m-2) = 6,$$
or
$$(2n-3)(m-1) = 3.$$
Therefore, either $m=3, n=9$ or $m=4, n=6$ or $m=5, n=5$ or $m=8, n=4$ or $m=2, n=3$ or $m=4, n=2$.
Note that only the case m = 2, n = 3 satisfies the following equation:
$$\frac{3}{n} + \frac{2}{m} = n^3 - 4n^2 + 4n - m^2 - m + 5.$$
Thus, it follows that $k = -7$.
Alternative solution. Apply the *rational root* theorem.   □

Problem 4.24. *Let ABCD be a cyclic quadrilateral, such that $AB = 3, BC = 8, CD = 6$ and $AD = 4$. Let M and N be points on line BD, such that AM is parallel to CD and CN is parallel to AB. What is the value of $AM + CN$?*

(A) 24   (B) 25.5   (C) 18   (D) 27   (E) 13.5

Solution. Answer. (E)
Note that $\angle AMD = \angle CDM = \angle BAC$ and $\angle ACB = \angle ADB$. Therefore $\triangle AMD \sim \triangle BAC$.
Hence, we obtain that
$$\frac{AM}{AB} = \frac{AD}{BC}.$$
Thus, it follows that $AM = 1.5$.

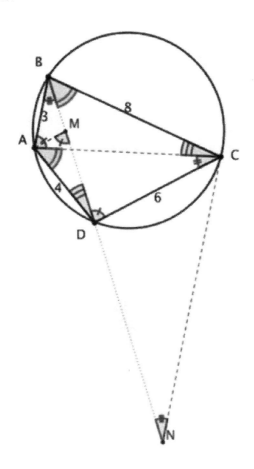

In a similar way, we obtain that $\angle CNB = \angle ABD = \angle ACD$ and $\angle NBC = \angle CAD$.
We deduce that $\triangle CBN \sim \triangle DAC$.
Thus, it follows that
$$\frac{CN}{CD} = \frac{BC}{AD}.$$
Hence, we obtain that $CN = 12$.
Therefore $AM + CN = 13.5$.   □

Problem 4.25. *Consider the set*

$$V = \{-3+2i, 2+i, -1\frac{1}{3}-3i\}.$$

Let $S = z_1 + z_2 + ... + z_{10}$, *where complex number* z_j *is an element of* V *for* $j = 1, 2, ..., 10$. *What is the probability that* $S = 0$?

(A) $\dfrac{280}{6561}$ (B) $\dfrac{140}{6561}$ (C) $\dfrac{280}{2187}$ (D) $\dfrac{140}{2187}$ (E) $\dfrac{70}{729}$

Solution. Answer. (A)

Let m, n, k be nonnegative integers and

$$\begin{cases} (-3+2i)m + (2+i)n + \left(-1\frac{1}{3}-3i\right)k = 0, \\ m+n+k = 10. \end{cases}$$

Then, we have that

$$\begin{cases} -3m + 2n - 1\frac{1}{3}k = 0, \\ 2m + n - 3k = 0, \\ m + n + k = 10. \end{cases}$$

Thus, it follows that $m = 2, n = 5, k = 3$. Hence, we obtain that

$$P(S=0) = \frac{\binom{10}{2} \cdot \binom{8}{5}}{3^{10}} = \frac{280}{6561}.$$

Therefore, the probability that $S = 0$ is equal to $\dfrac{280}{6561}$. $\square$

4.2 Solutions of AMC 12 type practice test 2

Problem 4.26. What is the value of the following expression?
$$(3^{-1} + 6^{-1} - 4^{-1})^{-1} : 5^0.$$

(A) $\frac{1}{4}$ (B) 4 (C) 5 (D) 1 (E) $\frac{4}{5}$

Solution. Answer. (B)
Note that
$$(3^{-1} + 6^{-1} - 4^{-1})^{-1} : 5^0 = (\frac{1}{3} + \frac{1}{6} - \frac{1}{4})^{-1} : 1 = 4.$$
$\square$

Problem 4.27. *Let the ratio of the perimeter of a triangle to its largest side be a positive integer less than 4. Given that its largest side is equal to 24. What is the value of the perimeter of the triangle?*

(A) 72 (B) 60 (C) 56 (D) 48 (E) 50

Solution. Answer. (A)
Let a, b, c be the side lengths of given triangle, such that $a \leq b \leq c = 24$. Given that
$$\frac{a+b+c}{c} < 4,$$
and by the triangle inequality $a + b > c$. Thus, it follows that
$$2 < \frac{a+b+c}{c} < 4.$$
Hence, as the ratio of the perimeter of a triangle to its largest side is a positive integer (less than 4 and greater than 2), we obtain that
$$\frac{a+b+c}{c} = 3.$$
We deduce that
$$a + b + c = 3c = 72.$$
Therefore, the perimeter of the triangle is equal to 72. $\square$

Problem 4.28. *There are 7 boys and 8 girls in a class. The teacher gave each of these 15 students a test with 10 problems. Each problem is scored either 1 or 0 points. Given that the average score of all girls was 7 and the average score of the entire class was 5.6. What was the average score of all boys?*

(A) 4.2 (B) 5 (C) 4 (D) 6 (E) 8

Solution. Answer. (C)
The total score of the class was $15 \cdot 5.6 = 84$, while the total score of all girls was $8 \cdot 7 = 56$.
Thus, it follows that the total score of all boys was $84 - 56 = 28$.
Therefore, the average score of all boys was $\frac{28}{7} = 4$. $\square$

Problem 4.29. *Let the ratio of the sum of the cubes of two positive numbers to the difference of their cubes be equal to $\dfrac{189}{61}$. By what percent is the larger number greater than the smaller number?*

(A) 50   (B) 25   (C) 100   (D) 40   (E) 60

Solution. Answer. (B)
Let x, y be considered positive numbers. Given that
$$\frac{x^3 + y^3}{x^3 - y^3} = \frac{189}{61}.$$
Thus, it follows that
$$128x^3 = 250y^3.$$
Hence, we obtain that
$$x = \frac{5}{4}y.$$
We deduce that the larger number is greater than the smaller number by the following percent:
$$\frac{\frac{5}{4}y - y}{y} \cdot 100 = 25.$$

□

Problem 4.30. *Given that $2 < a < 3, 3 < b < 4$ and $1 < c < 2$. Which of the following intervals contains the value of the expression $a - \dfrac{b}{c}$?*

(A) (-4,-3)   (B) (1.5, 2)   (C) (2, 3)   (D) (3, 5)   (E) (-2, 1.5)

Solution. Answer. (E)
We have that
$$\frac{1}{2} < \frac{1}{c} < 1$$
and $3 < b < 4$. Thus, it follows that
$$1.5 < \frac{b}{c} < 4.$$
Hence, we obtain that
$$-4 < -\frac{b}{c} < -1.5,$$
as well as $2 < a < 3$. We deduce that
$$-2 < a - \frac{b}{c} < 1.5.$$
Therefore, interval $(-2, 1.5)$ contains the value of the expression $a - \dfrac{b}{c}$. □

Problem 4.31. *Five years ago Henry was three times Mia's age. Given that (now) the sum of their ages is equal to 34. How old is Mia?*

(A) 6   (B) 10   (C) 13   (D) 11   (E) 8

Solution. Answer. (D)
Assume that (now) Henry is x years old and Mia is y years old. Given that $x + y = 34$ and $x - 5 = 3(y-5)$. Thus, it follows that
$$3(y-5) + 5 + y = 34.$$
Hence, we obtain that $y = 11$.
Therefore, now Mia is 11 years old. □

Problem 4.32. Let points A, B lie in the first quadrant of the coordinate plane and belong to the graph of function $y = \dfrac{1}{x^3}$. The abscissa of point A is 25% greater than the abscissa of point B. By what percent is the ordinate of point A smaller than the ordinate of point B?

(A) 75   (B) 25   (C) 20   (D) 50   (E) 48.8

Solution. Answer. (E)

Let x_0 be the abscissa of point B. Therefore, the abscissa of point A is $\dfrac{5}{4} x_0$.

The ordinates of points A and B are $\dfrac{64}{125 x_0^3}$ and $\dfrac{1}{x_0^3}$.

Therefore, the ordinate of point A is smaller than the ordinate of point B by the following percent:

$$\dfrac{1 - \dfrac{64}{125}}{1} \cdot 100 = 48.8.$$

□

Problem 4.33. Let the ratio of the lengths of the legs of a right triangle be $5 : 12$. Given that the diameter of its incircle is equal to D and the area of given triangle is equal to $m \cdot D^2$. What is the value of m?

(A) $\dfrac{15}{4}$ (B) $\dfrac{15}{8}$ (C) $\dfrac{25}{144}$ (D) $\dfrac{12}{5}$ (E) $\dfrac{5}{12}$

Solution. Answer. (B)

Let the lengths of the legs of given right triangle be equal to $5x$ and $12x$.
Then, using the Pythagorean theorem, we obtain that the hypothenuse of given right triangle is:

$$\sqrt{25x^2 + 144x^2} = 13x.$$

The inradius r of given triangle is:

$$r = \dfrac{5x + 12x - 13x}{2} = 2x.$$

Thus, it follows that $D = 4x$. Hence, we obtain that $x = \dfrac{D}{4}$.

The area of given triangle is:

$$\dfrac{5x \cdot 12x}{2} = 30x^2 = \dfrac{15}{8} D^2.$$

Therefore $m = \dfrac{15}{8}$. □

Problem 4.34. Carol randomly selects any two of the following numbers: 1, 2, 3, 4, 5, 6. Then, Claudia randomly selects any two numbers from the list of the remaining numbers. Finally, Cheryl selects the last two remaining numbers. What is the probability that one of the selected numbers of each girl is the multiple of the other selected number of the same girl?

(A) $\dfrac{1}{15}$ (B) $\dfrac{1}{4}$ (C) $\dfrac{1}{6}$ (D) $\dfrac{1}{10}$ (E) $\dfrac{1}{5}$

Solution. Answer. (A)

Note that one of them must choose the pair 1, 5 the other one must choose the pair 3, 6 and the third one must choose the pair 2, 4. Therefore, the probability that one of the selected numbers of each girl is the multiple of the other selected number of the same girl is:

$$\dfrac{3!}{\binom{6}{2} \cdot \binom{4}{2} \cdot \binom{2}{2}} = \dfrac{1}{15}.$$

□

Problem 4.35. For how many positive integers n the value of $\dfrac{20n}{n+15}$ is an integer?

(A) 18   (B) 10   (C) 9   (D) 7   (E) 8

Solution. Answer. (C)

Note that
$$\frac{20n}{n+15} = 20 - \frac{300}{n+15}.$$
Given that $\dfrac{300}{n+15}$ is an integer. Thus, the number $n+15$ is a divisor of 300 and it is greater than 15.
We have that $300 = 2^2 \cdot 3 \cdot 5^2$. Therefore, according to the subsection 1.3.2 (*number of divisors of a composite number*), we obtain that 300 has 18 divisors.
Note that from these 18 divisors 1, 2, 3, 4, 5, 6, 10, 12, 15 are not greater than 15.
Hence, the total number of divisors of 300 which are greater than 15 is equal to 9.
Therefore, for 9 positive integers n the value of $\dfrac{20n}{n+15}$ is an integer. □

Problem 4.36. *For any lines a, b, c in the plane let k be the number of all circles in that plane, such that each of them touches each of the lines a, b, c (note that the value of k depends on the choice of the lines a, b, c). How many possible values of k are there?*

(A) 0   (B) 1   (C) 4   (D) 3   (E) 2

Solution. Answer. (D)

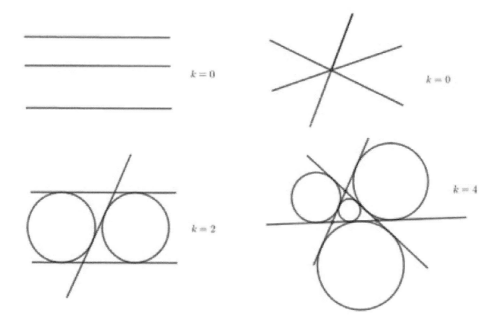

Note that k can be 0, 2 or 4. Therefore, there are 3 possible values of k. □

Problem 4.37. *Two parabolas given by equations $y = x^2 + bx - 2$ and $y = -3x^2 + (b-4)x + 6$, intersect each other at exactly two points on the x-axis. What is the value of the area of a quadrilateral which has vertices at these intersection points and at the vertices of the parabolas.*

(A) 27   (B) 13.5   (C) 13   (D) 12   (E) 10

Solution. Answer. (B)
Let the abscissas of the intersection points of the parabolas be x_1 and x_2. According to Vieta's formula, we have that
$$x_1 + x_2 = -\frac{b}{1} = \frac{b-4}{3}.$$
Thus, it follows that $b = 1$.
For $b = 1$, we have that $x_1 = 1$ and $x_2 = -2$.
Hence, the intersection points with the x-axis have coordinates $(-2,0)$ and $(1,0)$. The coordinates of the vertices are $\left(-\frac{1}{2}, -\frac{9}{4}\right)$ and $\left(-\frac{1}{2}, \frac{27}{4}\right)$.
Therefore, the area of the quadrilateral with vertices at these four points is $\frac{3 \cdot 9}{2} = 13.5$. □

Problem 4.38. *More than one boy and more than one girl attend math lessons. During the first lesson, each girl and each boy exchanged with each other one of their photos. During the second lesson, every two students of the same gender exchanged with each other one of their photos. Given that the total number of exchanges during the first lesson was the same as the total number of exchanges during the second lesson. How many students attend math lessons?*

(A) 3   (B) 5   (C) 7   (D) 4   (E) 10

Solution. Answer. (D)
Let the number of boys attending math lessons be m and the number of girls attending math lessons be n. Given that
$$mn = m(m-1) + n(n-1),$$
where $m \geq 2, n \geq 2$. We have that
$$(m-n)^2 + m\left(\frac{n}{2} - 1\right) + n\left(\frac{m}{2} - 1\right) = 0.$$
Thus, it follows that $m = n = 2$. Hence $m + n = 4$.
Therefore 4 students attend math lessons. □

Problem 4.39. *What is the value of a for which the following equation holds true?*
$$\log_2 3 \cdot \log_3 4 + \log_4 5 \cdot \log_5 6 \cdot \log_6 7 \cdot \log_7 a = 3.5.$$

(A) 4   (B) 5   (C) 6   (D) 7   (E) 8

Solution. Answer. (E)
We have that
$$\frac{\log 3}{\log 2} \cdot \frac{\log 4}{\log 3} + \frac{\log 5}{\log 4} \cdot \frac{\log 6}{\log 5} \cdot \frac{\log 7}{\log 6} \cdot \frac{\log a}{\log 7} = 3.5.$$
Hence, we obtain that
$$\frac{\log a}{\log 4} = 1.5.$$
Thus, it follows that $\log a = \log 4^{1.5}$. Therefore $a = 8$. □

Problem 4.40. *What is the smallest positive value of n for which the number $\frac{n}{10!}$ can be expressed as a finite decimal?*

(A) 405   (B) 189   (C) 567   (D) 135   (E) 199

Solution. Answer. (C)

Given that the number $\frac{n}{10!}$ can be expressed as a finite decimal, thus

$$\frac{n}{10!} = \frac{m}{10^k},$$

where $m > 0, k \geq 0$ and numbers m, k are integers. Hence, we obtain that

$$m = \frac{10^k n}{2^8 \cdot 3^4 \cdot 5^2 \cdot 7}.$$

Thus, it follows that $3^4 \cdot 7 \mid n$.

Therefore, the smallest positive value of n is 567.

Note that

$$\frac{567}{10!} = \frac{1}{2^8 \cdot 5^2} = \frac{5^6}{10^8} = 0,00015625.$$

$\square$

Problem 4.41. *Let $ABCD$ be a tetrahedron, such that $AB = 8, AC = 4, AD = 4, BC = \sqrt{34}, BD = 4\sqrt{3}$ and $CD = 5$. What is the volume of the tetrahedron?*

(A) $\frac{128}{3}$ (B) $\sqrt{39}$ (C) $\frac{128}{6}$ (D) $8\sqrt{34}$ (E) $2\sqrt{39}$

Solution. Answer. (E)

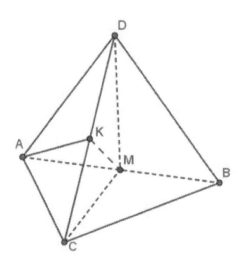

Let M be the midpoint of segment AB. From the formula of the median for triangles ABC, ABD and medians CM, DM we have that

$$CM = \sqrt{\frac{2 \cdot 4^2 + 2 \cdot (\sqrt{34})^2 - 8^2}{4}} = 3,$$

and
$$DM = \sqrt{\frac{2 \cdot 4^2 + 2 \cdot (4\sqrt{3})^2 - 8^2}{4}} = 4.$$

Note that
$$CM^2 + DM^2 = 3^2 + 4^2 = CD^2.$$

Thus, it follows that $\angle CMD = 90°$.

Let us calculate the volume of tetrahedron $AMCD$.

We have that $AM = AC = AD$. If K is the midpoint of segment CD, then we obtain that $AK \perp CD$ and $\triangle AKM \sim \triangle AKC$.

Hence, we deduce that $AK \perp MK$. Therefore, AK is the height of tetrahedron $AMCD$.

Thus, it follows that
$$Volume(AMCD) = \frac{1}{3} \cdot \frac{3 \cdot 4}{2} \cdot \sqrt{4^2 - \left(\frac{5}{2}\right)^2} = \sqrt{39}.$$

As M is the midpoint of segment AB and belongs to plane (MCD), then points A and B are equidistant from that plane. Thus, it follows that
$$Volume(ABCD) = 2\sqrt{39}.$$

$\square$

Problem 4.42. *Let a twenty-digit number starts with 2017 and its other digits are chosen randomly from the digits 0, 1, 2, 3, 4, 5, 6. What is the probability that such twenty-digit number is divisible by 140?*

(A) $\dfrac{18}{343}$ (B) $\dfrac{15}{343}$ (C) $\dfrac{12}{343}$ (D) $\dfrac{1}{49}$ (E) $\dfrac{4}{343}$

Solution. Answer. (E)

Let us enumerate the digits from the leftmost to the rightmost by $1^{st}, 2^{nd}, ..., 20^{th}$.

We have to choose the digits from 5th to 20th.

Note that a number is divisible by 140, if it is divisible by 5, 4 and 7. Obviously, the rightmost 20th digit must be 0 and 19th digit must be one of the numbers from 0, 2, 4, 6. Note that we can randomly choose the digits from 5th to 17th, while 18th digit is chosen uniquely. This is due to the fact that for any positive number M numbers $100 \cdot 0 + M, 100 \cdot 1 + M, 100 \cdot 2 + M, 100 \cdot 3 + M, 100 \cdot 4 + M, 100 \cdot 5 + M, 100 \cdot 6 + M$ leave different remainders after division by 7.

Hence, only one of these numbers is divisible by 7.

Therefore, the probability that such twenty-digit number is divisible by 140 is:
$$\frac{1}{7} \cdot \frac{4}{7} \cdot \frac{1}{7} = \frac{4}{343}.$$

$\square$

Problem 4.43. *For how many values of a in the interval $(-1, 1)$ does the quadratic expression $x^2 + ax + 3a + 2$ have at least one integer root?*

(A) 2 (B) 5 (C) 101 (D) 0 (E) 1

Solution. Answer. (A)

Let $a \in (-1, 1)$ and m be an integer root of the quadratic equation
$$x^2 + ax + 3a + 2 = 0.$$

Thus, it follows that
$$a = -\frac{m^2 + 2}{m + 3},$$

and
$$\left|\frac{m^2+2}{m+3}\right| < 1.$$

Thus $m^2 + 2 < |m + 3|$. Therefore $m^2 + 2 < |m| + 3$. We deduce that
$$(2|m| - 1)^2 < 5.$$

Hence, we obtain that $m \in \{0, 1\}$. Thus, it follows that
$$a \in \{-\frac{3}{4}, -\frac{2}{3}\}.$$

Therefore, for 2 values of a in the interval $(-1, 1)$ the quadratic expression $x^2 + ax + 3a + 2$ has at least one integer root. □

Problem 4.44. *Let $ABCD$ be a convex quadrilateral, such that $AB = 1, BC = 4, CD = 8, AD = 7$. What is the greatest possible value of the area of the quadrilateral?*

(A) $2\sqrt{65}$ (B) 17   (C) 18   (D) 19   (E) 18.5

Solution. Answer. (C)
Let us consider point B_1 symmetric to point B with respect to the perpendicular bisector of segment AC.

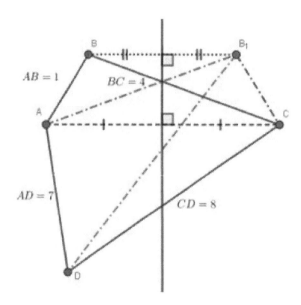

Note that
$$Area(\triangle ABC) = Area(\triangle AB_1C).$$

Thus, it follows that
$$Area(ABCD) = Area(AB_1CD) = Area(\triangle AB_1D) + Area(\triangle B_1CD) \le$$
$$\le \frac{1}{2}AB_1 \cdot AD + \frac{1}{2}B_1C \cdot CD = 14 + 4 = 18.$$

Now, let us consider quadrilateral AB_1CD, where $AB_1 = 4, B_1C = 1, CD = 8, AD = 7$ and $B_1D = \sqrt{65}$. Note that $Area(AB_1CD) = 18$ and $Area(ABCD) = 18$, where B is the symmetric to point B_1 with respect to the perpendicular bisector of segment AC. □

Problem 4.45. *Given a trapezoid with bases BC and AD, such that $AB = 4, BD = 3, AD = 5$ and $BC = 3$. What is the total number of all possible values of CD?*

(A) 1   (B) 3   (C) 4   (D) 2   (E) 5

Solution. Answer. (D)

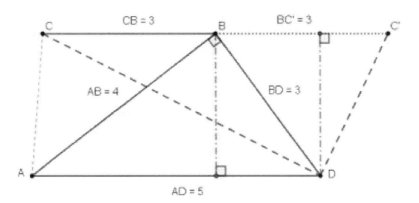

Note that triangle ABD and triangle BCD have equal heights (see the figure). Thus, it follows that
$$\frac{Area(ABD)}{Area(BCD)} = \frac{5}{3}.$$

Let $CD = 2x$, according to Heron's formula $Area(\triangle ABD) = 6$ and
$$Area(\triangle BCD) = \sqrt{(3+x) \cdot x \cdot x \cdot (3-x)}.$$

Hence, we obtain that
$$\frac{6}{x\sqrt{9-x^2}} = \frac{5}{3}.$$

Thus, it follows that
$$25x^4 - 225x^2 + 324 = 0.$$
We deduce that either $x = \sqrt{7.2}$ or $x = \sqrt{1.8}$.
Therefore the total number of all possible values of CD is equal to 2. $\square$

Problem 4.46. *Let F_1 and F_2 be the foci of the ellipse given by equation $\frac{x^2}{25} + \frac{y^2}{9} = 1$. Let C be a point on the ellipse, such that $\angle F_1CF_2 = 90°$. What is the value of the area of triangle F_1CF_2?*

(A) 16   (B) 9   (C) 10   (D) 12   (E) 12.5

Solution. Answer. (B)
Let $F_1(-c, 0)$ and $F_2(c, 0)$ be foci of the ellipse, where $c > 0$ (see the figure).

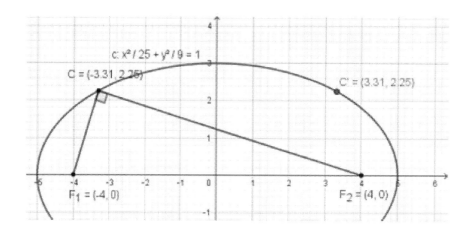

We have that $c^2 = 25 - 9 = 16$. For point $C(x_0, y_0)$ we have that $\angle F_1 C F_2 = 90°$. Hence, we obtain that
$$F_1 F_2^2 = F_1 C^2 + F_2 C^2.$$

Therefore
$$64 = (x_0 + 4)^2 + y_0^2 + (x_0 - 4)^2 + y_0^2.$$

Thus, it follows that
$$x_0^2 + y_0^2 = 16.$$

We also have that
$$\frac{x_0^2}{25} + \frac{y_0^2}{9} = 1.$$

We deduce that
$$\frac{16 - y_0^2}{25} + \frac{y_0^2}{9} = 1.$$

Therefore $|y_0| = \frac{9}{4}$. Thus, it follows that
$$Area(\triangle F_1 C F_2) = \frac{8 \cdot \frac{9}{4}}{2} = 9.$$

$\square$

Problem 4.47. *We say that a sequence of 12 numbers is "nice" if its each term is either 1, 2 or 3, and if the the sum of any four consecutive terms is either 7 or 9. How many "nice" sequences are there?*

(A) 100   (B) 36   (C) 256   (D) 128   (E) 472

Solution. Answer. (E)
Let $x_1, x_2, ..., x_{12}$ be a "nice" sequence.
Note that either three of the numbers x_1, x_2, x_3, x_4 are equal to 2, or only one of the numbers x_1, x_2, x_3, x_4 is equal to 2.
Beside that, if $i \leq 8$ and $x_i = 2$, then $x_{i+4} = 2$.
Let us consider the following two cases:
Case 1. If three of the numbers x_1, x_2, x_3, x_4 are equal to 2, then the total number of such sequences is $4 \cdot 2^3 = 32$.
Case 2. If only one of the numbers x_1, x_2, x_3, x_4 is equal to 2, then in the sequence $x_1, x_2, ..., x_{12}$ there will be exactly three twos (as if $i \leq 8$ and $x_i = 2$, then $x_{i+4} = 2$). Removing all twos from the sequence $x_1, x_2, ..., x_{12}$ we obtain a sequence of nine numbers each term of which is either 1 or 3. Moreover, any

three consecutive terms cannot be pariwise equal. Let us denote by a_n the total number of such sequences having n terms. Therefore, if only one of the numbers x_1, x_2, x_3, x_4 is equal to 2, then the total number of $x_1, x_2, ..., x_{12}$ "nice" sequences is equal to $4 \cdot a_9$.

We have that $a_n = y_n + z_n$, where y_n is the total number of sequences ending with 1 and z_n is the number of sequences ending with 3. If we replace the sequence $b_1, ..., b_n$ by the sequence $4 - b_1, ..., 4 - b_n$, then we obtain that $y_n = z_n$.

On the other hand, we have that $y_n = z_{n-1} + z_{n-2}, n = 3, 4, ...$

Thus, it follows that

$$a_n = 2y_n = 2y_{n-1} + 2y_{n-2} = a_{n-1} + a_{n-2}, n = 3, 4, ...$$

We have that $a_1 = 2, a_2 = 4$. Hence, we obtain that

$$a_3 = 6, a_4 = 10, a_5 = 16, a_6 = 26, a_7 = 42, a_8 = 68, a_9 = 110.$$

Therefore, in total there are $32 + 4 \cdot 110 = 472$ "nice" sequences. $\square$

Problem 4.48. *Let ABC be an equilateral triangle with a side length of 2. Let points M and N be randomly chosen points on line segments AB and AC, respectively. What is the probability that $MN \leq \sqrt{3}$?*

(A) $\dfrac{\pi}{6}$ (B) $\dfrac{1}{2}$ (C) $\dfrac{1}{3}$ (D) $\dfrac{\pi\sqrt{3}}{6}$ (E) $\dfrac{6-\pi}{\pi}$

Solution. Answer. (D)

Let $MA = u$ and $NA = v$. According to the law of cosines, we have that

$$MN^2 = u^2 - uv + v^2.$$

Thus, it follows that

$$u^2 - uv + v^2 \leq 3.$$

We deduce that

$$\left(u - \frac{v}{2}\right)^2 + \left(\frac{v\sqrt{3}}{2}\right)^2 \leq 3.$$

Let us denote

$$u - \frac{v}{2} = x,$$

and

$$\frac{v\sqrt{3}}{2} = y,$$

where $0 \leq u \leq 2$ and $0 \leq v \leq 2$. Thus, it follows that $0 \leq y \leq \sqrt{3}$ and

$$0 \leq x + \frac{y}{\sqrt{3}} \leq 2,$$

as well as

$$x^2 + y^2 \leq 3.$$

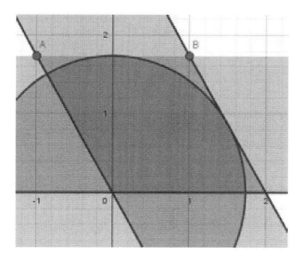

Therefore, the probability that $MN \leq \sqrt{3}$ is:

$$\frac{\frac{1}{3}\pi(\sqrt{3})^2}{2\sqrt{3}} = \frac{\pi\sqrt{3}}{6}$$

$\square$

Problem 4.49. *Let r and s be randomly chosen rational numbers from the interval $(0, 1)$. Each of them can be expressed in the form $\frac{n}{12}$, where n is an integer. What is the probability that the absolute value of the number $\cos(\pi r) + i\cos(\pi s)$ is equal to 1?*

(A) $\dfrac{19}{121}$ (B) $\dfrac{2}{11}$ (C) $\dfrac{20}{121}$ (D) $\dfrac{24}{121}$ (E) $\dfrac{28}{121}$

Solution. Answer. (C)
Given that
$$|\cos(\pi r) + i\cos(\pi s)| = 1.$$

Thus, it follows that
$$\cos^2(\pi r) + \cos^2(\pi s) = 1 = \cos^2(\pi r) + \sin^2(\pi r).$$

Hence, we obtain that
$$\cos(2\pi s) + \cos(2\pi r) = 0,$$

or
$$\cos\pi(s+r) \cdot \cos\pi(s-r) = 0.$$

Therefore, either $s + r = \dfrac{1}{2} + m$ or $s - r = \dfrac{1}{2} + m$, where m is an integer.
We deduce that, either $s + r = \dfrac{1}{2}$ or $s + r = \dfrac{3}{2}$ or $s - r = -\dfrac{1}{2}$ or $s - r = \dfrac{1}{2}$.
The total number of pairs (r, s) is $11 \cdot 11 = 121$.
The total number of pairs satisfying the above mentioned conditions is $4 \cdot 5 = 20$.
Hence, we deduce that the probability that the absolute value of the number $\cos(\pi r) + i\cos(\pi s)$ is equal to 1 is $\dfrac{20}{121}$.

$\square$

Problem 4.50. *Let parabola given by equation $y = x^2$ be drawn on a rectangular coordinate plane. Given a circle with the following properties: it passes through point (1, 1), it is tangent to the x−axis at a point that has a positive abscissa, the circle and the parabola have a common tangent line passing through point (1, 1). What is the radius of that circle?*

(A) $\dfrac{5 - \sqrt{5}}{4}$ (B) $\dfrac{1}{2}$ (C) $\dfrac{2}{3}$ (D) $\dfrac{5 - \sqrt{5}}{2}$ (E) $\dfrac{3 - \sqrt{5}}{2}$

Solution. Answer. (A)

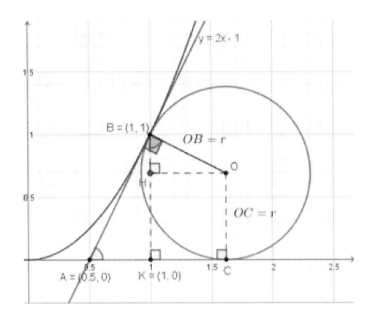

The equation of the tangent line of parabola $y = x^2$ passing through point $(1,1)$ is $y - 1 = 2(x - 1)$. Thus, it follows that $y = 2x - 1$.

Note that $\triangle ABK$ and $\triangle BHO$ are similar triangles, therefore
$$\frac{BH}{BO} = \frac{AK}{AB}.$$

Denoting the radius of the circle by r, we obtain that
$$\frac{1 - r}{r} = \frac{\frac{1}{2}}{\frac{\sqrt{5}}{2}}.$$

Hence, we deduce that
$$r = \frac{5 - \sqrt{5}}{4}.$$

□

4.3 Solutions of AMC 12 type practice test 3

Problem 4.51. *Let ABCD be a square with a side length of 6. Let E be a point on the line segment BC. What is the value of the area of triangle AED?*

(A) 30 (B) 24 (C) 12 (D) 18 (E) 6

Solution. Answer. (D)
Note that perpendicular drawn from point E on side AD has a length of 6.
Thus, it follows that
$$Area(\triangle AED) = \frac{1}{2} \cdot 6 \cdot 6 = 18.$$
□

Problem 4.52. *Among ten children on a hiking tour, any two of them have different number of candies. They split up equally into two groups, and it turns out that the total number of candies in the first group is five times smaller than the total amount of candies in the second group. What is the smallest possible total number of candies all children together can have?*

(A) 70 (B) 50 (C) 60 (D) 40 (E) 30

Solution. Answer. (C)
Note that the total number of candies in the first group is greater than or equal to $0+1+2+3+4 = 10$ and the total number of candies all children have is greater than or equal to $10 + 10 \cdot 5 = 60$.
Let us provide an example, such that all together they have 60 candies and the number of candy in both groups satisfy the assumptions of the problem: for the first group les us take $\{0, 1, 2, 3, 4\}$ and for the second group let us take $\{7, 8, 9, 10, 16\}$.
Therefore, the smallest possible total number of candies all children together can have is equal to 60. □

Problem 4.53. *There are 112 apples in one box, 97 apples in a second box, and 88 apples in a third box. First, a apples were transferred from the first box to the second box, and then b apples were transferred from the second box to the third box. After that, all the boxes had an equal number of apples. What is the value of $a + b$?*

(A) 11 (B) 13 (C) 20 (D) 25 (E) 24

Solution. Answer. (E)
As the total number of apples in all three boxes is equal (after the last transfer of apples from the second box to the third box), then the total number of apples in each box is $\frac{112 + 97 + 88}{3} = 99$.
Hence, we obtain that $a = 112 - 99 = 13$ and $b = 97 + 13 - 99 = 11$.
Thus, it follows that $a + b = 13 + 11 = 24$. □

Problem 4.54. *Given that $2^{2013} : 2^x = 2^{25} \cdot 2^{1975}$. What is the value of x?*

(A) 0 (B) $\frac{2013}{2010}$ (C) 4013 (D) 13 (E) -13

Solution. Answer. (D)
We have that
$$2^{2013-x} = 2^{25+1975}.$$

Thus, it follows that $2013 - x = 2000$. Hence, we deduce that $x = 13$. □

Problem 4.55. *Let in the classroom 22 students be seated in two rows. Given that 25% of the students in the first row and 20% of the students in the second row are enrolled in an after-school math program. How many students in the classroom are enrolled in the after-school math program?*

(A) 4   (B) 2   (C) 5   (D) 3   (E) 6

Solution. Answer. (C)
Let the number of students in the first row be n and the number of students in the second row be m. We have that $n + m = 22$.
Given that $\dfrac{n}{4}$ students in the classroom and $\dfrac{m}{5}$ students in the classroom are after-school math program students.
Therefore $4 \mid n$ and $5 \mid m$. Hence, we obtain that $n = 12$ and $m = 10$.
Thus, it follows that
$$\frac{n}{4} + \frac{m}{5} = 5.$$
Therefore, 5 students in the classroom are enrolled in the after-school math program. □

Problem 4.56. *Mary read a book in four days. The number of pages she read on the second day was two times less than the number of pages she read on the first day, the number of pages she read on the fourth day was 50% of the number of pages she read on the second day. Given that the ratio of the number of pages she read on the third day to the number of pages she read on the fourth day is $3:4$. What percent of the book did Mary read on the first day?*

(A) $\dfrac{20}{3}$ (B) 40   (C) 70   (D) $\dfrac{1600}{31}$ (E) 30.3

Solution. Answer. (D)
Let the number of pages that Mary has read on the first day be a. Thus, it follows that on the second day she has read $\dfrac{a}{2}$ pages, on the fourth day $\dfrac{a}{4}$ pages and on the third day
$$\frac{3}{4} \cdot \frac{a}{4} = \frac{3a}{16}$$
pages. Hence, on the first day she has read the following percent of the book:
$$\frac{a}{a + \dfrac{a}{2} + \dfrac{a}{4} + \dfrac{3a}{16}} \cdot 100 = \frac{1600}{31}.$$
□

Problem 4.57. *Let $x_1, x_2, ..., x_{20}$ be a sequence of numbers. Given that $x_1 = 2$ and*
$$x_{n+1} = \frac{n+2}{n+1} x_n,$$
for $n = 1, 2, ..., 19$. What is the value of x_{20}?

(A) 21   (B) 20   (C) $\dfrac{21}{20}$ (D) $\dfrac{20}{21}$ (E) $\dfrac{21}{16}$

Solution. Answer. (A)
We have that
$$x_2 = \frac{3}{2} x_1 = 3,\ x_3 = \frac{4}{3} x_2 = 4,\ ...,\ x_{20} = \frac{21}{20} x_{19} = 21.$$
Thus, it follows that $x_{20} = 21$. □

Problem 4.58. *Let x and y be different numbers, such that $20x + 13x^2 = 20y + 13y^2$. What is the value of $x + y$?*

(A) $\dfrac{13}{20}$ (B) $-\dfrac{20}{13}$ (C) $-\dfrac{13}{20}$ (D) $\dfrac{20}{13}$ (E) 0

Solution. Answer. (B)
We have that
$$(x - y)(20 + 13(x + y)) = 0.$$
Thus, it follows that
$$x + y = -\dfrac{20}{13}.$$

Problem 4.59. *Let a straight line containing incenter I of $\triangle ABC$ be parallel to side AC. Let M, N be intersection points of this line with sides AB, BC, respectively. Given that $MN = 10$ and $AC = 15$. What is the value of the perimeter of trapezoid $AMNC$?*

(A) 35   (B) 40   (C) 30   (D) 50   (E) 55

Solution. Answer. (A)

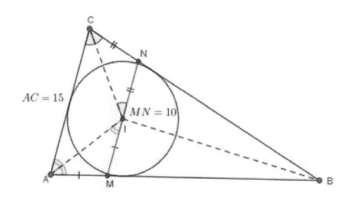

As I is the incenter of triangle ABC, then $\angle IAB = \angle IAC$ and $MN \parallel AC$.
Thus, it follows that
$$\angle MAI = \angle IAB = \angle IAC = \angle AIM.$$
Hence, we obtain that $AM = MI$.
In a similar way, we obtain that $CN = NI$. Therefore, the perimeter of trapezoid $AMNC$ is:
$$AM + MN + NC + AC = MI + MN + NI + AC = 2MN + AC = 35.$$

Problem 4.60. *Let a, b, c, d be digits. What is the greatest possible positive value of integer n satisying the following equation?*
$$\dfrac{1}{n} = 0.\overline{abcd} = 0.abcdabcd....$$

(A) 1111   (B) 909   (C) 303   (D) 9999   (E) 3333

Solution. Answer. (D)
Note that
$$\frac{10000}{n} - \overline{abcd} = \frac{1}{n}.$$
Hence, we obtain that $n \mid 9999$. Therefore, the greatest possible value of n is 9999 and we have that $\frac{1}{9999} = 0.\overline{(0001)}$. □

Problem 4.61. *Let AB and CD be the bases of trapezoid $ABCD$ (see the figure). Let M, N and P, K be points on bases AB and CD, respectively, such that $AB + CD = 18$ and*
$$Area(AMPD) = Area(MNKP) = Area(NBCK).$$
What is the length of the midsegment of trapezoid $MNKP$?

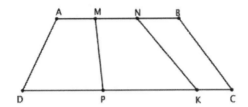

(A) 2.5   (B) 4.5   (C) 6   (D) 9   (E) 3

Solution. Answer. (E)
Let us denote by h the height of trapezoid $ABCD$. We have that
$$\frac{AM + PD}{2} h = \frac{MN + PK}{2} h = \frac{NB + CK}{2} h.$$
Thus, it follows that
$$AM + PD = MN + PK = NB + CK.$$
We deduce that
$$MN + PK = \frac{1}{3}(AB + CD) = 6.$$
Therefore, the length of the midsegment of trapezoid $MNKP$ is:
$$\frac{MN + PK}{2} = 3.$$
□

Problem 4.62. *Let the angle measures of a triangle with sides a, b, c form an arithmetic sequence. Given that $a \leq b \leq c$ and that $a, \sqrt{2}b, 3c$ form a geometric sequence. What is the value of the common ratio of this geometric sequence?*

(A) $\sqrt{2}$ (B) $\sqrt{3}$ (C) $\sqrt{6}$ (D) 2   (E) 3

Solution. Answer. (C)
Let the angles of given triangle be α, β, γ and $\alpha \leq \beta \leq \gamma$. Given that
$$\beta = \frac{\alpha + \gamma}{2},$$
and
$$\alpha + \beta + \gamma = 180°.$$

Thus, it follows that $\beta = 60°$. From the law of cosines, we obtain that
$$b^2 = a^2 + c^2 - 2ac \cos \beta.$$
We have that $2b^2 = 3ac$. Hence, we deduce that
$$a^2 + c^2 - 2.5ac = 0.$$
Therefore $c = 2a$. Thus, it follows that
$$\frac{\sqrt{2}b}{a} = \frac{\sqrt{6}a}{a} = \sqrt{6}.$$

□

Problem 4.63. *Let ABCD be a convex quadrilateral of area 18 and with vertex coordinates $A(1,1), B(2,7), C(m,n), D(6,3)$ on the xy–coordinate plane. What is the value of $m + n$?*

(A) 12 (B) 7 (C) 8 (D) 11 (E) 10

Solution. Answer. (D)

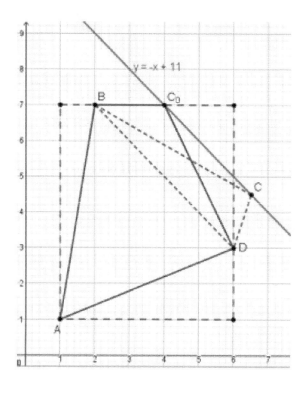

We have that the coordinates of C_0 are $(4, 7)$. Note that
$$Area(ABC_0D) = 5 \cdot 6 - \frac{6 \cdot 1}{2} - \frac{5 \cdot 2}{2} - \frac{4 \cdot 2}{2} = 18 = Area(ABCD).$$

Thus, it follows that $CC_0 \parallel BD$ and the equation of line BD is $y = -x + 9$.
Therefore, line CC_0 is "above" line BD by 2. Hence, the equation of line CC_0 is $y = -x + 11$.
Hence, we obtain that $m + n = 11$.

□

Problem 4.64. *Given that* $2^{\sqrt[4]{\log_2 3}} = 3^{\sqrt[4]{\log_3^3 x}}$. *What is the value of x?*

(A) $\sqrt{2}$ (B) 2   (C) $\sqrt[3]{2}$ (D) 4   (E) 3

Solution. Answer. (B)
Note that
$$2^{\sqrt[4]{\log_2 3}} = (3^{\log_3 2})^{\sqrt[4]{\log_2 3}} = 3^{\sqrt[4]{\log_3^4 2 \cdot \log_2 3}} = 3^{\sqrt[4]{\log_3^3 2}}.$$

Thus, it follows that
$$3^{\sqrt[4]{\log_3^3 2}} = 3^{\sqrt[4]{\log_3^3 x}}.$$

Hence, we obtain that
$$\sqrt[4]{\log_3^3 2} = \sqrt[4]{\log_3^3 x}.$$

We deduce that
$$\log_3 x = \log_3 2.$$

Therefore $x = 2$. □

Problem 4.65. *The principal of a school wants to give three identical math books, two identical physics books, and two identical chemistry books as gifts to the best four 12th graders. Each of the students should receive at least one of the books, but not two identical ones. In how many different ways can the principal give out the books?*

(A) 108   (B) 144   (C) 72   (D) 120   (E) 100

Solution. Answer. (A)
Note that the following two cases are possible.
Case 1. Out of 3 students with math books, only one of them received a physics book. In total there are $4 \cdot 3 \cdot 6 = 72$ such ways.
Case 2. Out of 3 students with math books, two of them received a physics book. In total there are $4 \cdot 3 \cdot 3 = 36$ such ways.
Therefore, the principal can give out the books in 108 different ways, as $72 + 36 = 108$. □

Problem 4.66. *There are three types of solutions with $10\%, 20\%, 30\%$ concentrations of salt, respectively. Given that if we mix the first and the second types, then the mixture has 16% concentration of salt. Given also that if we mix the second and the third types, then the mixture has 24% concentration of salt. What concentration of salt would the mixture of the first and the third types have?*

(A) 15   (B) 25   (C) 20   (D) 5   (E) 40

Solution. Answer. (C)
Let the masses of the first, second, third mixtures be x, y, z, respectively. Given that
$$\frac{10x}{100} + \frac{20y}{100} = \frac{16(x+y)}{100},$$
and
$$\frac{20y}{100} + \frac{30z}{100} = \frac{24(y+z)}{100}.$$
Thus, it follows that $x = \frac{2}{3}y$ and $z = \frac{2}{3}y$.
Let the mixture concentration of the first and third solutions be $p\%$. Hence, we obtain that
$$\frac{10x}{100} + \frac{30z}{100} = \frac{p(x+z)}{100}.$$

Therefore $p = 20$. □

Problem 4.67. *On the first day of his work a painter painted $\frac{1}{3}$ part of the wall. On the second day, he painted $\frac{1}{5}$ of the remaining part of the wall, and so on. On the nth day, he painted $\frac{1}{2n+1}$ of the remaining part of the wall. What part of the wall the painter still needed to paint after the sixth day?*

(A) $\frac{1}{3}$ (B) $\frac{1}{125}$ (C) $\frac{512}{3003}$ (D) $\frac{1024}{3003}$ (E) $\frac{2048}{3003}$

Solution. Answer. (D)

On the first day the painter painted $\frac{1}{3}$ part of the wall, so $\frac{2}{3}$ part of the wall remained.

On the second day the painter painted $\frac{1}{5}$ of the remaining part of the wall, therefore he painted $\frac{1}{5} \cdot \frac{2}{3}$ part of the wall and it remained to paint $\frac{2}{3} \cdot \frac{4}{5}$ part of the wall, and so on.

Therefore, on the sixth day it remained to paint the following part of the wall:
$$\frac{2}{3} \cdot \frac{4}{5} \cdot \frac{6}{7} \cdot \frac{8}{9} \cdot \frac{10}{11} \cdot \frac{12}{13} = \frac{1024}{3003}.$$

□

Problem 4.68. *The centers of eight congruent spheres of radii 1 are vertices of a cube with an edge length of 3. What is the radius of a sphere that is externally tangent to the eight given spheres?*

(A) $1.5\sqrt{3}$ (B) 1 (C) $1.5\sqrt{3} - 1$ (D) $1.5\sqrt{3} + 1$ (E) 3

Solution. Answer. (C)

Consider the following diagonal cross section of the cube (see the figure).

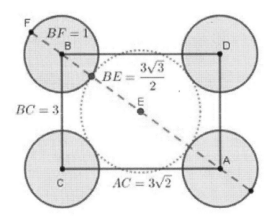

Note that according to the Pythagorean theorem the length of the diagonal of the face of the cube is equal to $3\sqrt{2}$ and the length of the main diagonal of the cube is equal to $3\sqrt{3}$, as $\sqrt{3^2 + (3\sqrt{2})^2} = 3\sqrt{3}$. Let us denote the radius of the sphere that is externally tangent to the eight given spheres by r and its center by E. Note that the distance from E to each of the vertices of the cube is the same and is equal to $r+1$, therefore E is the circumcenter of the cube. Note that the circumradius of the cube is equal to the half of the length of the main diagonal. Thus, it follows that
$$r + 1 = EB = \frac{3\sqrt{3}}{2}.$$

Hence, we obtain that
$$r = 1.5\sqrt{3} - 1.$$

□

Problem 4.69. Let a circle passing through vertices A, C of triangle ABC intersects sides AB, BC at points M, N, respectively. Given that the side lengths of triangle BMN are integers and $AB = 8, BC = 6$. What is the greatest possible length of line segment MN?

(A) 5 (B) 4 (C) 8 (D) 7 (E) 6

Solution. Answer. (E)
From the *power of a point* theorem it follows that $BM \cdot AB = BN \cdot BC$ (see the figure).
We obtain that $3BN = 4BM$. As the lengths of BN and BM are integers, then $BN = 4, BM = 3$.

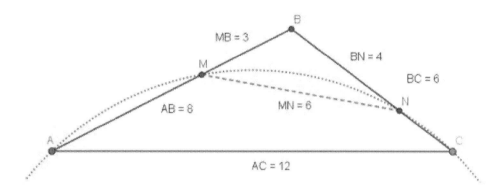

From triangle MBN according to the triangle inequality we have that $BM + BN > MN$.
As $BM + BN = 7$, therefore the greatest possible integer length of line segment MN is equal to 6.
Now, let us provide an example of a triangle satisfying the assumptions of the problem, such that $MN = 6$.
Consider a triangle ABC, such that $AB = 8, BC = 6, AC = 12$ and $BM = 3, BN = 4$.
In this case, note that triangles ABC and MBN are similar, as

$$\frac{BC}{MB} = \frac{AB}{BN}.$$

Thus, it follows that $\angle BAC = \angle MNB$ and $\angle ACB = \angle BNM$. Hence points A, M, N, C lie on the same circle. As triangles ABC and MBN are similar, then we have that

$$\frac{BC}{MB} = \frac{AC}{MN}.$$

Taking this into consideration and that $AC = 12$, we obtain that $MN = 6$. □

Problem 4.70. Consider the set $M = \{-14, -13, ..., -5, 1, 2, ..., 9\}$. How many ordered triples (a, b, c) exist, such that $a + b + c = 0$, where a, b, c are elements of M?

(A) 201 (B) 195 (C) 192 (D) 222 (E) 102

Solution. Answer. (B)
First note that, if a triple (a, b, c) satisfies the condition of the problem, then one of them is a negative number, while the other two are positive numbers.
Let us consider the following two cases.
Case 1. Two positive numbers from M are equal to each other. Then, positive numbers can be $\{3, 4, 5, 6, 7\}$ and there are 3 permutations of the relevant triple. Hence, in total there are $3 \cdot 5 = 15$ triples.
Case 2. Two positive numbers from M are not equal to each other. Then, the total number of triples is $6 \cdot (6 + 7 + 6 + 5 + 4 + 2) = 180$.
Thus, in total there are $15 + 180 = 195$ such triples. □

Problem 4.71. *What is the value of the integer part of the following expression?*

$$\log_8 9 + \log_9 10 + \ldots + \log_{63} 64.$$

(A) 56   (B) 60   (C) 57   (D) 59   (E) 53

Solution. Answer. (A)
Note that
$$\log_n(n+1) > \log_n n = 1,$$
where $n = 8, 9, \ldots, 63$. Thus, it follows that
$$\log_8 9 + \log_9 10 + \ldots + \log_{63} 64 > 56.$$

On the other hand, we have that

$$\log_8 9 + \log_9 10 + \ldots + \log_{63} 64 = 56 + \log_8\left(\frac{9}{8}\right) + \log_9\left(\frac{10}{9}\right) + \ldots + \log_{63}\left(\frac{64}{63}\right) <$$

$$< 56 + \log_8\left(\frac{9}{8}\right) + \log_8\left(\frac{10}{9}\right) + \ldots + \log_8\left(\frac{64}{63}\right) = 56 + \log_8\left(\frac{9}{8} \cdot \frac{10}{9} \cdot \ldots \cdot \frac{64}{63}\right) = 57.$$

Therefore, the integer part of $\log_8 9 + \log_9 10 + \ldots + \log_{63} 64$ is equal to 56.   □

Problem 4.72. *Let M be the set of all six-digit numbers, such that all digits of a number are from the set $\{1, 3, 4, 5, 6, 7, 9\}$. What is the probability that a randomly chosen number from set M is divisible by 7?*

(A) $\frac{1}{2}$ (B) $\frac{1}{3}$ (C) $\frac{1}{7}$ (D) $\frac{1}{5}$ (E) $\frac{1}{4}$

Solution. Answer. (C)
Let $\overline{abcdef} \in M$. Note that for any a, b, c, d, e the digit f can be chosen from set M in a unique way, such that $\overline{abcde0} + f$ is divisible by 7. This is due to the fact that numbers 1, 3, 4, 5, 6, 7, 9 leave different remainders after division by 7.
Thus, the probability that a randomly chosen number from set M is divisible by 7 is:

$$\frac{7 \cdot 7 \cdot 7 \cdot 7 \cdot 7 \cdot 1}{7 \cdot 7 \cdot 7 \cdot 7 \cdot 7 \cdot 7} = \frac{1}{7}.$$

□

Problem 4.73. *Given two congruent squares with a side length of 4, such that their centers coincide (see the figure). What is the smallest possible value of the area of the common interior part of these squares?*

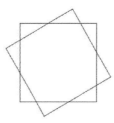

(A) 8   (B) $32(\sqrt{2} - 1)$ (C) $8\sqrt{2}$ (D) $16(\sqrt{2} - 1)$ (E) $4(\sqrt{2} + 1)$

Solution. Answer. (B)
Note that given two congruent squares are tangent to the same circle with a radius of 2 (see the figure).

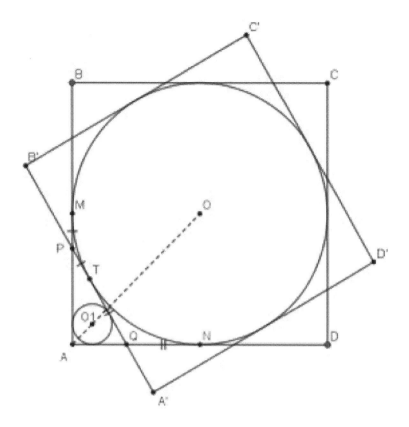

Let point O be the center of this circle. Note that after a rotation of $90°$ with respect to point O one of these squares coincides with the other square.
We have that the perimeter of triangle APQ is $AM + AN = 4$.
Thus, it follows that $Area(\triangle APQ) = 2r$, where r is its inradius.
On the other hand, we have that
$$AO_1 + O_1O = 2\sqrt{2}.$$
Hence, we deduce that
$$2\sqrt{2} \geq r\sqrt{2} + r + 2.$$
Thus, it follows that
$$r \leq 6 - 4\sqrt{2}.$$
Let us denote by S the area of the common interior part of these squares. We deduce that
$$S = 16 - 4 \cdot Area(\triangle APQ).$$
Thus, it follows that
$$S \geq 16 - 8(6 - 4\sqrt{2}) = 32(\sqrt{2} - 1).$$
Let us provide an example satisfying the assumptions of the problem, such that $S = 32(\sqrt{2} - 1)$.
If the second square is obtained from a rotation of $45°$ of the first square with respect to point O, then we have that
$$S = 32(\sqrt{2} - 1).$$
Therefore, the smallest possible value of the area of the common interior part of these squares is equal to $32(\sqrt{2} - 1)$.
□

Problem 4.74. *Four random vertices of a regular dodecagon (twelve-sided polygon) are chosen (see the figure). What is the probability that these four vertices form a quadrilateral with exactly two right angles?*

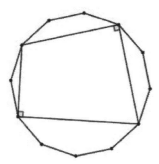

(A) $\dfrac{1}{4}$ (B) $\dfrac{1}{3}$ (C) $\dfrac{1}{6}$ (D) $\dfrac{8}{33}$ (E) $\dfrac{10}{33}$

Solution. Answer. (D)

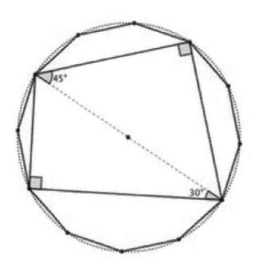

Note that we need to choose four vertices from given twelve vertices, thus the total number of all such quadrilaterals is equal to $\binom{12}{4}$.

It is known that a regular dodecagon (twelve-sided polygon) is cyclic (has a circumcircle). Hence, chosen quadrilateral has exactly two right angles, if its two vertices are chosen diametrically opposite to each other and the other two vertices are not diametrically opposite to each other.

Note that the total number of quadrilaterals, such that its two vertices are diametrically opposite to each other and the other two vertices are not diametrically opposite to each other is equal to $6 \cdot 5 \cdot 4$.

Therefore, the probability that chosen four vertices form a quadrilateral with exactly two right angles is:

$$\frac{6 \cdot 5 \cdot 4}{\binom{12}{4}} = \frac{8}{33}.$$

□

Problem 4.75. *How many real numbers a exist, such that the equation $z^3 + iz^2 + 5z + a = 0$ has a root of the form $z = x + iy$, where x and y are integers?*

(A) 0   (B) 2   (C) 4   (D) 3   (E) 1

Solution. Answer. (E)
Let $z = x + iy$, where x and y are integers. Thus, it follows that
$$(x+iy)^3 + i(x+iy)^2 + 5(x+iy) + a = 0.$$
We deduce that
$$x^3 - 3xy^2 - 2xy + 5x + a + i(3x^2y - y^3 + x^2 - y^2 + 5y) - 0.$$
Hence, we obtain that
$$\begin{cases} x^3 - 3xy^2 - 2xy + 5x + a = 0, \\ 3x^2y - y^3 + x^2 - y^2 + 5y = 0. \end{cases}$$
From the last equation, it follows that
$$x^2(3y+1) = y^3 + y^2 - 5y.$$
As y is an integer, therefore $3y + 1 \neq 0$. Hence, we obtain that
$$x^2 = \frac{y^3 + y^2 - 5y}{3y+1}.$$
In order to separate from the last fraction an expression with an integer value, let us multiple both sides of the last equation by 27. Thus, we have that
$$27x^2 = \frac{27y^3 + 27y^2 - 135y}{3y+1} = 9y^2 + 6y - 47 + \frac{47}{3y+1}.$$
As x and y are integers, then either $y = 0$ or $y = -16$.
If $y = -16$, then x is not an integer.
If $y = 0$, then $x = 0$.
Thus, it follows that $a = 0$. $\square$

4.4 Solutions of AMC 12 type practice test 4

Problem 4.76. *What is the value of the expression* $10 - (9 - (8 - (7 - (6 - (5 - (4 - (3 - (2 - 1))))))))$ *?*

(A) 1 (B) 3 (C) 0 (D) 7 (E) 5

Solution. Answer. (E)
Note that $10 - 9 + 8 - 7 + 6 - 5 + 4 - 3 + 2 - 1 = 5$.
Therefore, the value of the expression $10 - (9 - (8 - (7 - (6 - (5 - (4 - (3 - (2 - 1))))))))$ is equal to 5. □

Problem 4.77. *It took 5 hours to drive a car from city A to city B. In the first hour the car covered 50 miles and starting from the second hour it covered 10 miles per hour more than in each previous hour. What is the total distance (in miles) covered by the car?*

(A) 350 (B) 260 (C) 300 (D) 340 (E) 320

Solution. Answer. (A)
Given that in the first hour the car covered 50 miles and starting from the second hour it covered 10 miles per hour more than in each previous hour. As it took 5 hours to drive from city A to city B, then we have that the total distance covered by the car is $50 + 60 + 70 + 80 + 90 = 350$ (miles). □

Problem 4.78. *The area of the intersection part of two circles is 10% of the area of one of the circles, and 40% of the area of the other circle. What is the ratio of the radius of the larger circle to the radius of the smaller circle?*

(A) $\sqrt{2}$ (B) 4 (C) 2 (D) π (E) $\sqrt{3}$

Solution. Answer. (C)
Let the radii of the larger circle and the smaller circle be R and r, respectively. Given that

$$\frac{\pi r^2 \cdot 40}{100} = \frac{\pi R^2 \cdot 10}{100}.$$

Thus, it follows that

$$\frac{R}{r} = 2.$$

Therefore, the ratio of the radius of the larger circle to the radius of the smaller circle is equal to 2. □

Problem 4.79. *What is the range of the solutions of the following inequality?*

$$\frac{x}{2^x} + \sqrt[5]{x} + \frac{|x|}{x} \geq 0.$$

(A) $[0, +\infty)$ (B) $(0, +\infty)$ (C) $(-\infty, 0)$ (D) $[1, +\infty)$ (E) $(-\infty, +\infty)$

Solution. Answer. (B)
Note that $x = 0$ is not a solution of given inequality, because the denominator of a fraction cannot be equal to 0. On the other hand, we have that for any $x \neq 0$ each of the summands:

$$\frac{x}{2^x}, \sqrt[5]{x}, \frac{|x|}{x}$$

has the sign of x. Thus, it follows that the range of the solutions of given inequality is $(0, +\infty)$. □

Problem 4.80. *David solved 93 problems over the course of several days. Each day he solved either 3, 4 or 5 problems. Given that the total number of days during which he solved 3 or 4 problems is not more than 12, and the total number of days during which he solved 4 or 5 problems is not more than 17. What is the greatest number of days he could have spent solving 4 problems in a day?*

(A) 4 (B) 6 (C) 8 (D) 7 (E) 5

Solution. Answer. (D)
Let us denote by m, k, n the number of days that David has solved 3, 4, 5 problems, respectively. Given that
$$\begin{cases} 3m + 4k + 5n = 93, \\ m + k \leq 12, \\ k + n \leq 17. \end{cases}$$

Thus, it follows that
$$93 = 3m + 4k + 5n \leq 3(12 - k) + 4k + 5(17 - k) = 121 - 4k.$$

Hence, we obtain that
$$4k \leq 28.$$

We deduce that
$$k \leq 7.$$

Note that
$$k = 7, m = 5, n = 10$$
satisfy the assumptions of the problem.
Therefore, the greatest number of days he could have spent solving 4 problems in a day is equal to 7. □

Problem 4.81. *A palindrome number is a number that remains the same when its digits are reversed, for example 16061. What is the smallest possible positive difference of two different four-digit palindrome numbers?*

(A) 10 (B) 11 (C) 9 (D) 110 (E) 99

Solution. Answer. (B)
Let $\overline{abba}$ be a four-digit palindrome. Note that
$$\overline{abba} = 1001a + 110b = 11 \cdot (91a + 10b).$$

Hence, any four-digit palindrome number is divisible by 11.
Therefore, we deduce that the positive difference of two different palindrome numbers is divisible by 11 and is not equal to 0.
Note that 2002 and 1991 are palindrome numbers. We have that $2002 - 1991 = 11$.
Hence, we obtain that the smallest possible positive difference of two different four-digit palindrome numbers is equal to 11. □

Problem 4.82. *Given three spheres, the radii of the first two spheres are 3 and 4. The sum of the total surface area of these two spheres is equal to the total surface area of the third sphere. What is the value of the volume of the third sphere?*

(A) 125π (B) 200π (C) 100π (D) $\dfrac{400\pi}{3}$ (E) $\dfrac{500\pi}{3}$

Solution. Answer. (E)
Let the radius of the third sphere be R. Given that the sum of the total surface area of the first two spheres (of radii 3 and 4) is equal to the total surface area of the third sphere, therefore
$$4\pi \cdot 3^2 + 4\pi \cdot 4^2 = 4\pi R^2.$$
Thus, it follows that $R = 5$. Hence, the volume of the third sphere is:
$$\frac{4}{3}\pi \cdot R^3 = \frac{500\pi}{3}.$$

□

Problem 4.83. *Let the longest side of triangle ABC be AC. Let BH be an altitude, such that $AC = 4 \cdot BH$ and $\angle C = 15°$. What is the value of the angle measure (in degrees) of $\angle A$?*

(A) 15   (B) 30   (C) 75   (D) 45   (E) 60

Solution. Answer. (C)
Let us choose point M, such that
$$\angle MBC = \angle MCB = 15°.$$

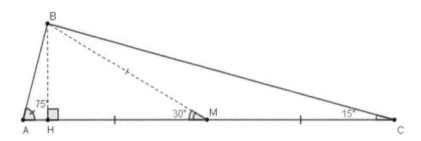

Thus, it follows that $\angle HMB = 30°$. Hence, we obtain that $\triangle HMB$ is a $30 - 60 - 90$ triangle. We deduce that
$$MB = 2 \cdot BH = \frac{AC}{2}.$$
On the other hand, we have that $MB = MC$. Therefore
$$MA = AC - MC = AC - MB = \frac{AC}{2}.$$
Thus, it follows that $MA = MB$. We obtain that
$$\angle A = \frac{180° - 30°}{2} = 75°.$$
Therefore, the value of the angle measure (in degrees) of $\angle A$ is 75.

□

Problem 4.84. Let the edge length of a cube be 4 inches. A right circular cylinder and a square prism are removed from this cube (see the figure). Given that the radius of the base of the cylinder is 1 inch and the centers of its bases coincide with the centers of two opposite faces of the cube. The centers of the bases of the square prism coincide with the centers of two other opposite faces of the cube. The base edge length of the square prism is 2 inches and all its lateral faces are parallel to the corresponding faces of the cube. What is the value of the volume (in cubic inches) of the remaining solid?

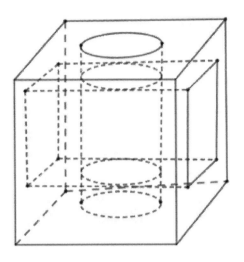

(A) $56 - 4\pi$ (B) $64 - 4\pi$ (C) 48 (D) 32 (E) $48 - 2\pi$

Solution. Answer. (E)
Note that the value of the volume (in cubic inches) of the remaining solid is:
$$4^3 - 2 \cdot 2 \cdot 4 - 2 \cdot \pi \cdot 1^2 \cdot 1 = 48 - 2\pi.$$

□

Problem 4.85. Let p and q be given numbers, such that $p - 2q, p - q, p + q$ and 40 form a geometric sequence. What is the smallest possible value of the sum of the first three terms of this sequence?

(A) 120 (B) 35 (C) 20 (D) 24 (E) 30

Solution. Answer. (B)
Given that $p - 2q, p - q, p + q$ form a geometric sequence, therefore
$$(p - q)^2 = (p - 2q)(p + q).$$
Thus, it follows that
$$q(p - 3q) = 0.$$
We deduce that either $q = 0$ or $p = 3q$.
If $q = 0$, then given geometric sequence is $p, p, p, 40$. Hence, it is $40, 40, 40, 40$.
If $p = 3q$, then given geometric sequence is $q, 2q, 4q, 40$. Hence, it is $5, 10, 20, 40$.
Therefore, the smallest value of the sum of the first three terms of this sequence is $5 + 10 + 20 = 35$. □

Problem 4.86. What is the value of the greatest solution of the equation $3^{x^2+x} = 4^{x+1}$?

(A) -1 (B) 0   (C) $2\log_3 2$ (D) $\log_4 3$ (E) $\log_3 2$

Solution. Answer. (C)
Given equation is equivalent to the following equation:
$$\left(\frac{3^x}{4}\right)^{x+1} = 1.$$
Thus, it follows that either $3^x = 4$ or $x + 1 = 0$.
Hence, we obtain that either $x = \log_3 4 = 2\log_3 2$ or $x = -1$.
Therefore, the value of the greatest solution of given equation is equal to $2\log_3 2$. □

Problem 4.87. *Each of the inhabitants of TruLi island $(A, B, C, D$ and $E)$ either always tells the truth or always lies. They once said the following:*
A said: "C is a truth-teller."
B said: "D is a truth-teller."
C said: "E is a truth-teller."
D said: "B is a truth-teller."
E said: "B is a liar."
What is the value of the product of the number of truth-tellers and the number of liars on TruLi island?

(A) 6   (B) 4   (C) 0   (D) 5   (E) 3

Solution. Answer. (A)
Let us consider the following two cases.
Case 1. B is a truth-teller, then as per B: D is a truth-teller, as per E: E is a liar, as per C: C is a liar, as per A: A is a liar. Therefore, there are 3 liars and 2 truth-tellers.
Case 2. B is a liar, then A, C, E are truth-tellers and B, D are liars.
Therefore, the answer is $3 \cdot 2 = 6$. □

Problem 4.88. *What is the greatest possible value of a, such that the following system of equations has a solution in the set of real numbers?*
$$\begin{cases} y = x^2 + 7x + 3a, \\ x = y^2 + 7y + \dfrac{1}{3}a^3. \end{cases}$$

(A) 9   (B) 6   (C) 7   (D) 3   (E) 2

Solution. Answer. (D)
Let (x_0, y_0) be a solution of given system of equations, then
$$y_0 = x_0^2 + 7x_0 + 3a$$
and
$$x_0 = y_0^2 + 7y_0 + \frac{1}{3}a^3.$$
Summing up these two equations, we obtain that
$$(x_0 + 3)^2 + (y_0 + 3)^2 = 18 - 3a - \frac{a^3}{3}.$$
Thus, it follows that
$$18 - 3a - \frac{a^3}{3} \geq 0.$$

We obtain that
$$(a-3)(a^2 + 3a + 18) \leq 0.$$
Therefore $a \leq 3$.
If $a = 3$, then $(-3, -3)$ is a solution of given system of equations.
Hence, the greatest possible value of a, such that given system of equations has a solution in the set of real numbers is equal to 3. $\square$

Problem 4.89. *Let D be a point on side AC of triangle ABC. Given that $AB = 6, AC = 9, CD = 5$ and $BD = 8$. What is the value of the length of side BC?*

(A) $5\dfrac{1}{3}$ (B) 10 (C) 8 (D) 12 (E) 14

Solution. Answer. (D)

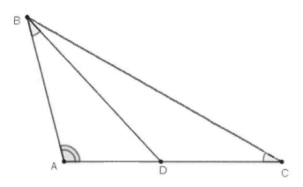

We have that $AD = AC - CD = 4$. Note that
$$\frac{AD}{AB} = \frac{AB}{AC}.$$
Thus, it follows that $\triangle ADB \sim \triangle ABC$. Therefore
$$\frac{BD}{BC} = \frac{4}{6}.$$
Hence, we obtain that $BC = 12$.
Alternative solution. As $AB = 6, AC = 9, CD = 5, BD = 8$ and $AD = 4$, then from Stewart's theorem, it follows that $BC = 12$. $\square$

Problem 4.90. *Let the probability that it will rain some day during a given week be equal to $\dfrac{1}{3}$. Which one of the following events has the greatest probability to happen?*

(A) It will rain exactly 2 days during a week.
(B) It will rain exactly 3 days during a week.
(C) It will rain exactly 4 days during a week.
(D) It will rain exactly 5 days during a week.
(E) It will rain exactly 6 days during a week.

Solution. Answer. (A)
The probability of raining exactly k days during a week is:
$$\binom{7}{k} \cdot \left(\frac{1}{3}\right)^k \left(\frac{2}{3}\right)^{7-k}.$$

Note that
$$\binom{7}{6}\left(\frac{1}{3}\right)^6 \cdot \frac{2}{3} < \binom{7}{5}\left(\frac{1}{3}\right)^5 \left(\frac{2}{3}\right)^2 < \binom{7}{4}\left(\frac{1}{3}\right)^4 \left(\frac{2}{3}\right)^3 <$$
$$< \binom{7}{3}\left(\frac{1}{3}\right)^3 \left(\frac{2}{3}\right)^4 < \binom{7}{2}\left(\frac{1}{3}\right)^2 \left(\frac{2}{3}\right)^5 > \binom{7}{1} \cdot \frac{1}{3} \cdot \left(\frac{2}{3}\right)^6.$$

Therefore, from given events the greatest probability to happen has event (A). □

Problem 4.91. *Ben randomly chooses a number from each of the sets: $\{1, 3, 5, 7, 9, 11\}$ and $\{2, 4, 6, 8, 10\}$. What is the probability that the sum of the chosen numbers is a multiple of 3?*

(A) $\frac{4}{15}$ (B) $\frac{1}{3}$ (C) $\frac{1}{5}$ (D) $\frac{1}{4}$ (E) $\frac{1}{6}$

Solution. Answer. (B)
Suppose Ben did choose numbers a and b from the first set (the first six odd numbers) and from the second set (the first five even numbers), respectively.
Therefore, the total number of all possible (a, b) pairs is $6 \cdot 5 = 30$.
On the other hand, the total number of favorable outcomes is equal to 10 (see the table below).

	1	3	5	7	9	11
2	3	5	7	9	11	13
4	5	7	9	11	13	15
6	7	9	11	13	15	17
8	9	11	13	15	17	19
10	11	13	15	17	19	21

Therefore, the probability that the sum of the chosen numbers is a multiple of 3 is $\frac{10}{30} = \frac{1}{3}$. □

Problem 4.92. *Let $ABCDEF$ be a convex hexagon, such that $\angle B = \angle D = \angle F = 120°$, $\angle C = 2 \cdot \angle ACE$ and $\angle A = 2 \cdot \angle CAE$. Given that the area of triangle ACE is equal to $10\sqrt{3}$ and the area of hexagon $ABCDEF$ is equal to $11\sqrt{3}$. What is the value of $|DE - EF|$?*

(A) $\sqrt{3}$ (B) 3 (C) $3\sqrt{3}$ (D) 6 (E) $6\sqrt{3}$

Solution. Answer. (D)
Note that
$$\angle E = 360° - \angle A - \angle C = 360° - 2\angle ACE - 2\angle CAE = 2\angle AEC.$$

Thus, it follows that
$$\angle E = 2\angle AEC.$$

Let us consider symmetric triangles of triangles ABC, CDE, AEF with respect to lines AC, CE, AE (see the figure).

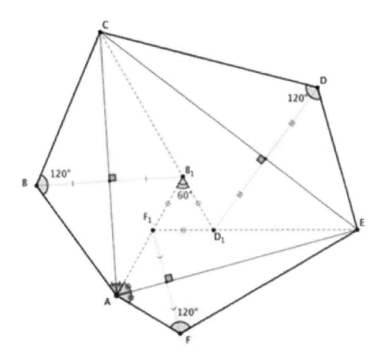

Note that $\triangle B_1 D_1 F_1$ is an equilateral triangle and its area is:
$$10\sqrt{3} - (11\sqrt{3} - 10\sqrt{3}) = 9\sqrt{3}.$$

On the other hand, its area is equal to:
$$\frac{\sqrt{3}}{4} F_1 D_1^2.$$

Therefore $F_1 D_1^2 = 36$. Thus, it follows that $F_1 D_1 = 6$. We deduce that
$$|DE - EF| = |D_1 E - EF_1| = F_1 D_1 = 6.$$

□

Problem 4.93. *The interior of a square with vertices (3, 1), (5, 3), (3, 5), (1,3) is cut out of the interior of a square with vertices (0, 0), (0, 6), (6, 6), (6, 0). An ant departs from (0, 0) and moves in the remaining interior of the bigger square, such that from each point (i, j) it moves straight either to the point $(i+1, j)$ or to the point $(i, j+1)$, where i and j are integers. Given that the ant stops at (6, 6). How many possible paths are there for the ant to reach (6, 6) from (0, 0)?*

(A) 51   (B) 102   (C) 124   (D) 50   (E) 200

Solution. Answer. (C)
In the figure below, every number written at each intersection point represents the number of possible paths from (0, 0) to that point.

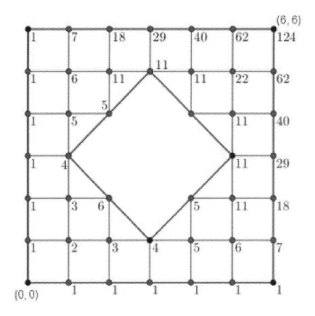

Therefore, there are 124 possible paths for the ant to reach (6, 6) from (0, 0). □

Problem 4.94. *Sam is randomly choosing n unit squares from different columns of $n \times n$ square grid. The probability that n chosen unit squares are in different columns and in different rows is $\frac{5}{324}$. What is the value of n?*

(A) 5   (B) 7   (C) 8   (D) 6   (E) 12

Solution. Answer. (D)

According to the multiplication rule of counting, in total there are n^n possibilities.
On the other hand, the total number of favorable outcomes is:
$$n(n-1)(n-2) \cdot \ldots \cdot 1 = n!.$$

Thus, it follows that
$$\frac{n!}{n^n} = \frac{(n-1)!}{n^{n-1}} = \frac{5}{324}.$$

If $n = 6$, then
$$\frac{(n-1)!}{n^{n-1}} = \frac{5!}{6^5} = \frac{5}{324}.$$

If $n \geq 7$, then
$$\frac{(n-1)!}{n^{n-1}} = \frac{5!}{n^5} \cdot \frac{6}{n} \cdot \ldots \cdot \frac{n-1}{n} < \frac{5!}{n^5} \cdot 1 \cdot \ldots \cdot 1 < \frac{5!}{6^5} = \frac{5}{324}.$$

Note that if $n \leq 5$, then
$$\frac{(n-1)!}{n^{n-1}} > \frac{5!}{6^5}.$$

Therefore $n = 6$. □

Problem 4.95. Let n be a positive integer and $(a_n), (b_n)$ be increasing arithmetic sequences with integer terms. Given that $a_1 = b_1 = 1$ and k is such a positive integer greater than 2 that $\dfrac{1}{a_k} + \dfrac{1}{b_k} = \dfrac{1}{10}$. What is the value of $\dfrac{1}{a_2} + \dfrac{1}{b_2}$?

(A) $\dfrac{1}{6}$ (B) $\dfrac{1}{3}$ (C) $\dfrac{1}{4}$ (D) $\dfrac{1}{2}$ (E) 1

Solution. Answer. (E)
From given equation, we deduce that
$$(a_k - 10)(b_k - 10) = 100.$$
As $(a_n), (b_n)$ are arithmetic sequences, then $k - 1 \mid a_k - a_1$ and $k - 1 \mid b_k - b_1$.
Thus, it follows that
$$(a_k - 1)(b_k - 1) - 9(a_k - 1) - 9(b_k - 1) = 19.$$
We deduce that $k - 1 \mid 19$ (and $k > 2$). Hence $k = 20$. Therefore $a_{20} \geq 20, b_{20} \geq 20$.
From the first equation, we obtain that $a_{20} = 20, b_{20} = 20$. Hence $a_2 = b_2 = 2$. Thus, it follows that
$$\frac{1}{a_2} + \frac{1}{b_2} = 1.$$

$\square$

Problem 4.96. Let $p(x)$ be a polynomial with integer coefficients. Given that $|p(x)| \leq 2010x^2$ for all real values of x. How many such polynomials exist?

(A) 4021   (B) 2010   (C) 3   (D) 2011   (E) 100

Solution. Answer. (A)
We have that $|p(0)| \leq 0$. Thus, it follows that $p(0) = 0$.
Therefore $p(x) = xq(x)$, where $q(x)$ is also a polynomial with integer coefficients. We deduce that
$$|xq(x)| \leq 2010x^2.$$
Hence, for all $x \neq 0$ we obtain that
$$|q(x)| \leq 2010|x|.$$
Let $x = \dfrac{1}{n}$, where n is a positive integer. Thus, it follows that
$$\left|q\left(\frac{1}{n}\right)\right| \leq \frac{2010}{n}.$$
When $n \to \infty$, we have that $|q(0)| \leq 0$.
Hence, we obtain that $q(x) = x \cdot s(x)$, where $s(x)$ is also a polynomial with integer coefficients. Moreover, for all $x \neq 0$ we have that
$$|s(x)| \leq 2010.$$
Let $s(x) = a_0 x^m + ... + a_m$, where $a_0, ..., a_m$ are integers.
If $m \geq 1$, then $a_0 \neq 0$ and
$$\left|a_0 + a_1 \cdot \frac{1}{n} + ... + a_m \cdot \frac{1}{n^m}\right| \leq \frac{2010}{n^m},$$
where n is a positive integer and when $n \to \infty$ we have that $|a_0| \leq 0$. This leads to a contradiction.
Therefore $s(x) = a_0$ and
$$|a_0| \leq 2010.$$

We obtain that
$$a_0 \in \{-2010, ..., 2010\}.$$
Note that all polynomials of the form $p(x) = a_0 x^2$ satisfy the assumptions of the problem, where $a_0 \in \{-2010, ..., 2010\}$. Therefore, there are 4021 such polynomials. $\square$

Problem 4.97. *Let x be a real number. What is the smallest possible value of the following expression?*
$$|x - \log_2 3| + |x - \log_3 4| + |x - \log_4 5|.$$

(A) 0 (B) 1 (C) $\log_2 \dfrac{3\sqrt{5}}{5}$ (D) $\log_2 3$ (E) 2

Solution. Answer. (C)

Let us prove that $\log_2 3 > \log_3 4 > \log_4 5$. We have that
$$\log_2 3 = \frac{\log 3}{\log 2}, \log_3 4 = \frac{\log 4}{\log 3}.$$

Note that
$$\log 2 \cdot \log 4 < \left(\frac{\log 2 + \log 4}{2}\right)^2 < \left(\frac{\log 9}{2}\right)^2 = \log^2 3.$$

Thus, it follows that $\log_2 3 > \log_3 4$.

In a similar way, we can prove that $\log_3 4 > \log_4 5$.

Let us denote given expression by $f(x)$, therefore
$$f(x) = |\log_2 3 - x| + |x - \log_4 5| + |x - \log_3 4| \geq |\log_2 3 - x| + |x - \log_4 5| \geq$$
$$\geq |(\log_2 3 - x) + (x - \log_4 5)| = \log_2 3 - \log_4 5.$$

On the other hand, we have that
$$f(\log_3 4) = |\log_2 3 - \log_3 4| + |\log_3 4 - \log_4 5| = \log_2 3 - \log_3 4 + \log_3 4 - \log_3 5 =$$
$$= \log_2 3 - \log_4 5.$$

Therefore, the smallest possible value of $f(x)$ is:
$$\log_2 3 - \log_4 5 = \log_2 3 - \log_2 \sqrt{5} = \log_2 \frac{3\sqrt{5}}{5}.$$
$\square$

Problem 4.98. *What is the sum of the last two digits of the sum $1^3 + 2^3 + ... + 2010^3$?*

(A) 10 (B) 7 (C) 9 (D) 5 (E) 17

Solution. Answer. (B)

We have that
$$a^3 + b^3 = (a+b)(a^2 - ab + b^2).$$

Thus, it follows that
$$100 \mid (-10)^3 + 2010^3, ..., 100 \mid 0^3 + 2000^3, 100 \mid 1^3 + 1999^3, 100 \mid 2^3 + 1998^3, ..., 100 \mid 999^3 + 1001^3.$$

Therefore, the last two digits of the sum $1^3 + 2^3 + ... + 2010^3$ coincide with the last two digits of the following sum:
$$1^3 + 2^3 + ... + 10^3 + 1000^3.$$

Note that this sum ends with 25. Therefore, the sum of the last two digits is equal to 7. $\square$

Problem 4.99. *What is the value of n for which the solution of the following inequality represents a union of n disjoint intervals?*

$$\log_x(2x) \cdot \log_{3x}(4x) \cdot \log_{5x}(6x) \cdot \log_{7x}(8x) \cdot \log_{9x}(10x) < 0.$$

(A) 10   (B) 4   (C) 6   (D) 3   (E) 5

Solution. Answer. (E)
Note that $\log_a b$ and $(a-1)(b-1)$ have the same sign.
Therefore, the given inequality is equivalent to the following system of inequalities:

$$\begin{cases} x > 0, \\ (x-1)(2x-1)(3x-1)(4x-1)(5x-1)(6x-1)(7x-1)(8x-1)(9x-1)(10x-1) < 0. \end{cases}$$

The solution of this system of inequalities is

$$\left(\frac{1}{10}, \frac{1}{9}\right) \cup \left(\frac{1}{8}, \frac{1}{7}\right) \cup \left(\frac{1}{6}, \frac{1}{5}\right) \cup \left(\frac{1}{4}, \frac{1}{3}\right) \cup \left(\frac{1}{2}, 1\right).$$

Thus, it follows that $n = 5$.   □

Problem 4.100. *Let M be the set $\{1, 2, ..., 16\}$. What is the total number of all three-element subsets of M, such that each of the subsets does not contain consecutive numbers?*

(A) 378   (B) 350   (C) 364   (D) 320   (E) 560

Solution. Answer. (C)
The total number of three-element subsets of M is $\binom{16}{3}$.
Now, let us find the total number of three-element subsets of M such that each of these subsets contains consecutive numbers.
Note that each of these subsets can contain either one pair of consecutive numbers or two pairs of consecutive numbers.
Note that the total number of subsets containing two pairs of consecutive numbers is equal to 14.
On the other hand, the total number of subsets containing one pair of consecutive numbers is $2 \cdot 13 + 13 \cdot 12 = 182$.
Therefore, the total number of all three-element subsets of M, such that each of the subsets does not contain consecutive numbers is $560 - 196 = 364$.
Alternative solution.
Let subset $\{a, b, c\}$ has no consecutive numbers.
Without loss of generality, one can assume that $a < b < c$.
Let us consider the following subset $\{a, b-1, c-2\}$. Note that

$$a, b-1, c-2 \in \{1, ..., 14\}.$$

Thus, it follows that there is one-to-one correspondence between the subsets containing no consecutive numbers and the set $\{1, 2, ..., 14\}$.
Therefore, total number of all three-element subsets of M, such that each of the subsets does not contain consecutive numbers is $\binom{14}{3} = 364$.   □

4.5 Solutions of AMC 12 type practice test 5

Problem 4.101. *A car left city A toward city B at 8:30 AM. The driver reached city B, rested for 25 minutes in city B and drove back to city A, reaching it at 1:05 PM. How long (in minutes) was the car driving on the road?*

(A) 275 (B) 225 (C) 240 (D) 250 (E) 295

Solution. Answer. (D)
Note that there is 4 hours and 35 minutes time difference between 8 : 30 AM and 1 : 05 PM. Therefore, the car was on the road for 4 hours and 10 minutes. That equals to 250 minutes. □

Problem 4.102. *Given that* $1 - \dfrac{1}{1 - \dfrac{1}{1+x}} = 2$. *What is the value of x?*

(A) -1 (B) 0 (C) $\dfrac{1}{3}$ (D) -0.5 (E) 0.5

Solution. Answer. (D)
We have that
$$1 - 2 = \dfrac{1}{1 - \dfrac{1}{1+x}}.$$

Thus, it follows that
$$1 - \dfrac{1}{1+x} = -1.$$

We obtain that
$$1 + x = \dfrac{1}{2}.$$

Therefore $x = -0.5$. □

Problem 4.103. *Five numbers are inserted between integers 1 and 2, such that all seven numbers together form an arithmetic sequence. What is the value of the smallest inserted number?*

(A) $\dfrac{6}{5}$ (B) $\dfrac{7}{6}$ (C) $\dfrac{4}{3}$ (D) $\dfrac{3}{2}$ (E) 1.1

Solution. Answer. (B)
Denote the terms of this arithmetic sequence by $a_1, a_2, a_3, a_4, a_5, a_6, a_7$, where $a_1 = 1$ and $a_7 = 2$. Thus, for the common difference d of this arithmetic sequence we have that:
$$d = \dfrac{2-1}{7-1} = \dfrac{1}{6}.$$

As the common difference is positive, thus the smallest inserted number is a_2. Hence
$$a_2 = 1 + \dfrac{1}{6} = \dfrac{7}{6}.$$

Therefore, the value of the smallest inserted number is equal to $\dfrac{7}{6}$. □

Problem 4.104. *There is $500 in each of three envelopes. The first envelope contains only $10 bills, the second one only $20 bills, and the third one only $50 bills. One, two, and three bills are taken out of these envelopes (in any order). What is the smallest possible amount (in dollars) that could have been taken out?*

(A) 130   (B) 120   (C) 110   (D) 230   (E) 100

Solution. Answer. (B)
Given that we need to take one, two and three bills of these envelopes (in any order). On the other hand, we have that $50 > 20 > 10$. Therefore, we obtain the smallest possible amount in (dollars) if we take one $50 bill, two $20 bills, three $10 bills.
Therefore, the smallest possible amount (in dollars) is $50 \cdot 1 + 20 \cdot 2 + 10 \cdot 3 = 120$. □

Problem 4.105. *If the width of a rectangle is increased by 2 units and the length is decreased by 2 units, then the area is equal to 28sq. units. On the other hand, if the width of the initial rectangle is decreased by 2 units and the length is increased by 2 units, then the area is equal to 24sq. units. What is the value (in sq. units) of the area of the initial rectangle?*

(A) 28   (B) 30   (C) 24   (D) 16   (E) 18

Solution. Answer. (B)
Let the width of the initial rectangle be a (units) and the length be b (units).
Given that
$$\begin{cases} (a+2)(b-2) = 28, \\ (a-2)(b+2) = 24. \end{cases}$$
Summing up these two equations, we obtain that $2ab - 8 = 52$. Hence $ab = 30$.
Therefore, the value (in sq. units) of the area of the initial rectangle is equal to 30. □

Problem 4.106. *Let m, n be integers, such that $2^m \cdot 3^n = a$ and $2^n \cdot 3^m = b$. What is the value of $3^{n^2 - m^2}$?*

(A) $a^n \cdot b^m$ (B) $a^m \cdot b^{-n}$ (C) $(ab)^{n-m}$ (D) $a^{-n} \cdot b^{-m}$ (E) $a^n \cdot b^{-m}$

Solution. Answer. (E)
Note that
$$2^{mn} \cdot 3^{n^2} = a^n,$$
and
$$2^{mn} \cdot 3^{m^2} = b^m.$$
Thus, it follows that
$$\frac{a^n}{b^m} = 3^{n^2 - m^2}.$$
Therefore, we obtain that
$$3^{n^2 - m^2} = a^n \cdot b^{-m}.$$
□

Problem 4.107. Let (b_n) be a geometric sequence with common ratio equal to 7. Given that the sum of its first five terms is equal to 41. What is the value of $b_6 + b_7 - b_1 - b_2$?

(A) 1968   (B) 2009   (C) 41   (D) 287   (E) 100

Solution. Answer. (A)
Given that
$$b_1 + b_2 + b_3 + b_4 + b_5 = 41.$$
Note that
$$b_3 + b_4 + b_5 + b_6 + b_7 = b_1 \cdot 7^2 + b_2 \cdot 7^2 + b_3 \cdot 7^2 + b_4 \cdot 7^2 + b_5 \cdot 7^2 = 41 \cdot 7^2 = 2009.$$
Note also that
$$b_6 + b_7 - b_1 - b_2 = (b_3 + b_4 + b_5 + b_6 + b_7) - (b_1 + b_2 + b_3 + b_4 + b_5) = 2009 - 41 = 1968.$$

□

Problem 4.108. *Given five unit squares (see the figure). What is the length (in units) of the side of the greatest square?*

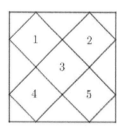

(A) 2   (B) 3   (C) $2\sqrt{2}$ (D) $\sqrt{7}$ (E) $\sqrt{6}$

Solution. Answer. (C)
Let the length of the side of the greatest square be a (units). Then, for its area we have that:
$$a^2 = 5 \cdot 1 + 4 \cdot \frac{1}{4} + 4 \cdot \frac{1}{2} = 8.$$
Thus, it follows that $a = 2\sqrt{2}$.
Therefore, the length (in units) of the side of the greatest square is equal to $2\sqrt{2}$.   □

Problem 4.109. Let $f(x) = ax^2 + bx + c$ and $f(x + 2030) = x^2 - x - 1$, for any value of x. What is the value of $b^2 - 4ac$?

(A) 2030   (B) −3   (C) 5   (D) 4   (E) 25

Solution. Answer. (C)
Given that $f(x) = ax^2 + bx + c$ and $f(x + 2030) = x^2 - x - 1$, therefore for any value of x we have that
$$a(x + 2030)^2 + b(x + 2030) + c = x^2 - x - 1.$$
We obtain that
$$ax^2 + (b + 4060a)x + 2030^2 a + 2030b + c = x^2 - x - 1.$$

140

Hence, we deduce that $a = 1$. On the other hand, for values x_0 and $x_0 - 2030$ quadratic functions $f(x)$ and $f(x + 2030)$ have the same values. Therefore, the minimum values of quadratic functions $f(x)$ and $f(x + 2030)$ are the same. Thus, it follows that

$$-\frac{D}{4a} = -\frac{(-1)^2 - 4 \cdot 1 \cdot (-1)}{4 \cdot 1}.$$

We obtain that
$$b^2 - 4ac = (-1)^2 - 4 \cdot 1 \cdot (-1) = 5.$$

□

Problem 4.110. *Let ABC be a triangle, such that $AB = 1$ and $BC = 12$. Given that the length (in units) of median BM is an integer number. What is the value of the length of BM?*

(A) 3   (B) 4   (C) 7   (D) 5   (E) 6

Solution. Answer. (E)
Let point N be the midpoint of side BC (see the figure). Thus, it follows that

$$MN = \frac{AB}{2} = 0.5,$$

$$BN = \frac{BC}{2} = 6.$$

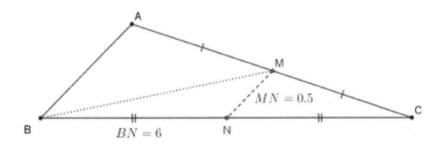

From $\triangle BMN$ according to the triangle inequality, we have that

$$BN - NM < BM < BN + NM.$$

Hence, we obtain that $5.5 < BM < 6.5$.
Given that the length (in units) of median BM is an integer number, therefore $BM = 6$.   □

Problem 4.111. *What is the value of the following expression?*

$$1^3 + 2 \cdot 2^2 + 2 \cdot 3^2 + ... + 2 \cdot 19^2 + 20^2 + 1 \cdot 2 + 2 \cdot 3 + 3 \cdot 4 + ... + 19 \cdot 20.$$

(A) 7999   (B) 8000   (C) 7200   (D) 789   (E) 800

Solution. Answer. (A)
Let us rewrite the given sum in the following way.

$$1^3 + 2 \cdot 2^2 + 2 \cdot 3^2 + ... + 2 \cdot 19^2 + 20^2 + 1 \cdot 2 + 2 \cdot 3 + 3 \cdot 4 + ... + 19 \cdot 20 =$$

$$= (2^2 + 2 \cdot 1 + 1^2) + (3^2 + 3 \cdot 2 + 2^2) + ... + (20^2 + 20 \cdot 19 + 19^2) = \frac{2^3 - 1^3}{2 - 1} + \frac{3^3 - 2^3}{3 - 2} + ... + \frac{20^3 - 19^3}{20 - 19} = 20^3 - 1 = 7999.$$

□

Problem 4.112. *How many positive integers have the following property: the integer is 6 less than the sum of the squares of all of its digits?*

(A) 1   (B) 3   (C) 2   (D) 0   (E) 6

Solution. Answer. (B)
Let $\overline{a_1 a_2 ... a_n}$ be a positive integer, such that
$$\overline{a_1 a_2 ... a_n} = a_1^2 + a_2^2 + ... + a_n^2 - 6.$$
If $n \geq 3$, then
$$\overline{a_1 a_2 ... a_n} = 10^{n-1} a_1 + 10^{n-2} a_2 + ... + 10 a_{n-1} + a_n \geq 100 a_1 + 10 a_2 + ... + 10 a_{n-1} + a_n \geq$$
$$\geq 10 a_1 + 10 a_2 + ... + 10 a_{n-1} + 90 a_1 > a_1^2 + a_2^2 + ... + a_{n-1}^2 + a_n^2 > a_1^2 + a_2^2 + ... + a_{n-1}^2 + a_n^2 - 6.$$
We deduce that
$$\overline{a_1 a_2 ... a_n} > a_1^2 + a_2^2 + ... + a_n^2 - 6.$$
This leads to a contradiction. Therefore $n < 3$.
We have to consider the cases $n = 1$ and $n = 2$.
If $n = 1$, then
$$a_1 = a_1^2 - 6.$$
Therefore $a_1 = 3$.
If $n = 2$, then
$$\overline{a_1 a_2} = a_1^2 + a_2^2 - 6.$$
Thus, it follows that
$$a_1(10 - a_1) + 6 = a_2(a_2 - 1).$$
One can easily verify (by simple casework) that $a_1 = 4, a_2 = 6$ or $a_1 = 6, a_2 = 6$.
Hence, there are 3 positive integers satisfying the assumptions of the problem. $\square$

Problem 4.113. *Let ABC and $A_1 B_1 C_1$ be triangles. Given that $AB = A_1 B_1 = 20$, $BC = B_1 C_1 = 10\sqrt{2}$ and $\angle ABC - \angle A_1 B_1 C_1 = 90°$. What is the greatest possible value of $AC - A_1 C_1$?*

(A) $10\sqrt{2}$ (B) 10   (C) 20   (D) 30   (E) $20\sqrt{2}$

Solution. Answer. (C)
At first, let us consider the following arrangement of $\triangle ABC$ and $\triangle A_1 B_1 C_1$ (see the figure).

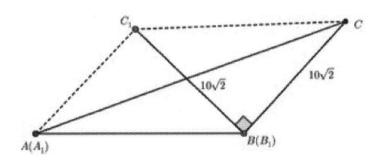

According to the triangle inequality, we have that
$$AC - AC_1 \leq CC_1 = 20.$$
Thus, it follows that
$$AC - A_1 C_1 \leq 20.$$

If C_1 lies on side AC (see the figure), then we have that
$$AC - A_1C_1 = 20.$$

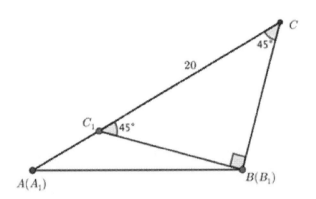

Therefore, the greatest possible value of $AC - A_1C_1$ is equal to 20. □

Problem 4.114. *On the xy–coordinate plane points (0,0), (0,3), (4,3) form a triangle. For what value of k does the line $y = kx + 1$ divide the perimeter of this triangle in two equal halfs?*

(A) $\dfrac{4}{3}$ (B) $\dfrac{3}{4}$ (C) 2   (D) $-\dfrac{1}{2}$ (E) $\dfrac{1}{2}$

Solution. Answer. (E)

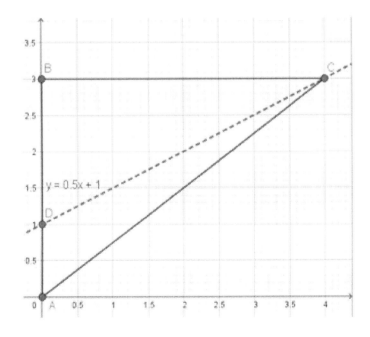

Note that the perimeter of triangle ABC is $3 + 4 + 5 = 12$ (see the figure). On the other hand, line $y = kx + 1$ passes through point $D = (0, 1)$. Therefore, it passes through point $C = (4, 3)$ too. Thus, it follows that
$$k = \frac{y-1}{4} = \frac{3-1}{4} = \frac{1}{2}.$$
□

Problem 4.115. *Let a, b be real numbers, such that $i + 2i^2 + \ldots + 2009i^{2009} = a + bi$. What is the value of $a + b$?*

(A) 1 (B) 2008 (C) 2010 (D) 2009 (E) -1

Solution. Answer. (D)
We have that
$$i + 2i^2 + \ldots + 2009i^{2009} = a + bi.$$
Multiplying both sides of this equation by i, we obtain that
$$i^2 + 2i^3 + \ldots + 2009i^{2010} = ai - b.$$
Subtracting the last equation from the first equation, we deduce that
$$i + i^2 + \ldots + i^{2009} - 2009i^{2010} = a + b + (b - a)i.$$
Thus, it follows that
$$\frac{i(i^{2009} - 1)}{i - 1} - 2009(i^2)^{1005} = a + b + (b - a)i.$$
Hence, we obtain that
$$i + 2009 = a + b + (b - a)i.$$
Therefore $a + b = 2009$. □

Problem 4.116. *Let two circles of radii 2 and 4 be externally tangent to each other. Given that line l is tangent to both circles and that both circles lie on the same side of line l. What is the radius of the circle that is externally tangent to both circles and with the center on line l?*

(A) $\dfrac{8}{7}$ (B) 1 (C) 1.5 (D) $\dfrac{7}{8}$ (E) $\dfrac{2}{3}$

Solution. Answer. (A)
Let $OD = OE = x$ (see the figure).

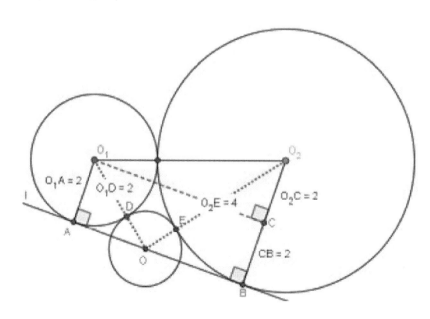

From right triangles OAO_1, OBO_2, O_1O_2C according to the Pythagorean theorem, we obtain that

$$\begin{cases} AO = \sqrt{x^2 + 4x}, \\ BO = \sqrt{x^2 + 8x}, \\ O_1C = \sqrt{6^2 - 2^2} = 4\sqrt{2}. \end{cases}$$

On the other hand, we have that $O_1C = AB = AO + OB$.
Thus, it follows that

$$\sqrt{x^2 + 8x} + \sqrt{x^2 + 4x} = 4\sqrt{2}.$$

Hence, we obtain that

$$4x = 4\sqrt{2}\left(\sqrt{x^2 + 8x} - \sqrt{x^2 + 4x}\right).$$

Summing up the last two equations, we deduce that

$$\sqrt{x^2 + 8x} = 2\sqrt{2} + \frac{x}{2\sqrt{2}}.$$

Thus, it follows that

$$\frac{7}{8}x^2 + 6x - 8 = 0.$$

Therefore $x = \dfrac{8}{7}$. □

Problem 4.117. *Let the sum of all the terms of the infinite geometric sequence* $1, \dfrac{1}{\sqrt[2009]{2}}, ...$ *be equal to* S. *What is the value of the sum of the infinite geometric sequence* $1, \dfrac{1}{S}, ...$?

(A) $\sqrt[2009]{4}$ (B) $\sqrt[2009]{2} + 1$ (C) 2 (D) $1 + \dfrac{1}{\sqrt[2009]{2}}$ (E) $\sqrt[2009]{2}$

Solution. Answer. (E)
Let
$$A = \frac{1}{\sqrt[2009]{2}}.$$

Given that
$$S = \frac{1}{1 - A}.$$

Thus, it follows that
$$S - 1 = \frac{A}{1 - A}.$$

Hence, we obtain that
$$\frac{S}{S - 1} = \frac{1}{A}.$$

The sum of the second infinite geometric sequence is

$$\frac{1}{1 - \frac{1}{S}} = \frac{S}{S - 1} = \frac{1}{A} = \sqrt[2009]{2}.$$

□

Problem 4.118. *Let n be a positive integer. How many terms of the sequence $x_n = 10^n - 3^n + 2^n + 5$ are perfect squares?*

(A) 0 (B) 2 (C) 1 (D) 4 (E) 3

Solution. Answer. (C)
Note that
$$3 \mid 2^n + 1,$$
when n is odd and
$$3 \mid 10^n - 1,$$
for any n. Therefore, if n is odd, then
$$x_n = (10^n - 1) - 3^n + (2^n + 1) + 5$$
leaves a remainder of 2 after division by 3. Hence, it cannot be a perfect square.
We have that
$$x_2 = 10^2 - 3^2 + 2^2 + 5 = 10^2$$
and when k is a positive integer greater than 1, then
$$(10^k - 1)^2 < 10^{2k} - 3^{2k} + 2^{2k} + 5 < 10^{2k},$$
as
$$9^k - 4^k = (9 - 4)(9^{k-1} + \ldots + 4^{k-1}) > 5.$$
Therefore, x_{2k} is in between two consecutive perfect squares. Hence, it cannot be a perfect square. Thus, it follows that the only possibility for x_n to be a perfect square is when $n = 2$. □

Problem 4.119. *What is the value of the greatest possible perimeter of a pentagon for which there exists an annulus with the area of 4π, such that given pentagon can be entirely instered into that annulus? Note that some (or all) vertices of given pentagon can also lie on the circumferences of two concentric circles of that annulus. (An annulus is a plane figure bounded by the circumferences of two concentric circles with different radii).*

(A) 10π (B) 20 (C) 10 (D) 20π (E) 15

Solution. Answer. (B)
Let a line segment of length a be inserted into a circular ring (annulus) with the area of 4π (see the figure).

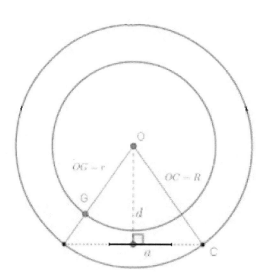

We have that
$$a \le 2\sqrt{R^2 - d^2} \le 2\sqrt{R^2 - r^2} = 2\sqrt{\frac{\pi R^2 - \pi r^2}{\pi}} = 2\sqrt{\frac{4\pi}{\pi}} = 4.$$

Thus, it follows that $a \le 4$.

Therefore, the perimeter of a convex pentagon that can be inserted into a circular ring (annulus) with the area of 4π is not greater than $5 \cdot 4 = 20$.

Note that a regular pentagon with a side of 4 can be inserted into a circular ring (annulus) with the area of 4π.

Therefore, the value of the greatest possible perimeter of a pentagon satisfying the assumptions of the problem is equal to 20. □

Problem 4.120. *Let the diagonals of a convex quadrilateral ABCD intersect at point O. Given that the areas of $\triangle ABO$ and $\triangle CDO$ are equal to 8 and 18, respectively. What is the smallest possible value of the area of ABCD?*

(A) 39   (B) 50   (C) 52   (D) 40   (E) $10\sqrt{5}$

Solution. Answer. (B)

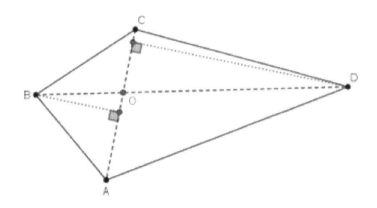

Note that triangles ABO and BOC have a common altitude drawn from vertex B (see the figure). Thus, it follows that
$$\frac{Area(ABO)}{Area(BOC)} = \frac{AO}{OC}.$$

In a similar way, we obtain that
$$\frac{Area(ADO)}{Area(DOC)} = \frac{AO}{OC}.$$

Hence, we deduce that
$$\frac{Area(ABO)}{Area(BOC)} = \frac{Area(ADO)}{Area(DOC)}.$$

Therefore $Area(ADO) \cdot Area(BOC) = 8 \cdot 18 = 144$.

Note that
$$Area(ABCD) = Area(ABO) + Area(BOC) + Area(CDO) + Area(ADO) =$$
$$= Area(ADO) + Area(BOC) + 26 = (\sqrt{Area(ADO)} - \sqrt{Area(BOC)})^2 + 50.$$

Thus, the smallest possible value of the area of $ABCD$ is 50 (sq. units).

An example of such quadrilateral (with area of 50 sq. units) is an isosceles trapezoid with bases AB, CD, such that its diagonals are perpendicular to each other and $AO = BO = 4, CO = DO = 6$. □

Problem 4.121. Let $P(x) = x^3 + ax^2 + bx + c$ be a polynomial with real coefficients. Given that both equations $P(x) = 3$ and $P(x) = 1$ have two different real roots. What is the value of $2a^3 - 9ab + 27c$?

(A) 25   (B) 37   (C) 42   (D) 54   (E) 63

Solution. Answer. (D)
Note that the polynomial
$$P'(x) = 3x^2 + 2ax + b$$
should have two real roots, otherwise $y = P(x)$ would be an increasing function. Therefore, both equations $P(x) = 3$ and $P(x) = 1$ would have only one real root.
Moreover, if x_1, x_2 are the roots of the equation $P'(x) = 0$, such that $x_1 < x_2$, then $P(x_1) = 3$ and $P(x_2) = 1$.
According to Vieta's formula, we have that
$$\begin{cases} x_1 + x_2 = -\dfrac{2a}{3}, \\ x_1 \cdot x_2 = \dfrac{b}{3}. \end{cases}$$
Moreover
$$x_1^3 + ax_1^2 + bx_1 + c = 3$$
and
$$x_2^3 + ax_2^2 + bx_2 + c = 1.$$
Summing up the last two equations, we obtain that
$$(x_1 + x_2)^3 - 3x_1x_2(x_1 + x_2) + a((x_1 + x_2)^2 - 2x_1x_2) + b(x_1 + x_2) + 2c = 4.$$
Thus, it follows that
$$2a^3 - 9ab + 27c = 54.$$
□

Problem 4.122. *Let all 8 edges of a regular square pyramid are equal to 1 unit. Given that a plane parallel to one of its lateral faces intersects the pyramid and the perimeter of the intersection part is equal to $2\dfrac{1}{3}$. What is the value (in sq. units) of the area of the intersection part?*

(A) $\dfrac{5\sqrt{3}}{36}$ (B) $\dfrac{\sqrt{3}}{8}$ (C) $\dfrac{\sqrt{3}}{16}$ (D) $\dfrac{4\sqrt{3}}{81}$ (E) $\dfrac{1}{64}$

Solution. Answer. (A)

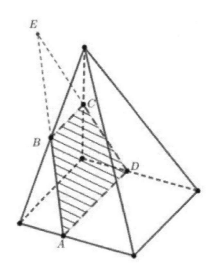

One can easily verify that the intersection is an isosceles trapezoid $ABCD$ (see the figure). Moreover AED is an equilateral triangle with a unit side.

Let $EB = x$. Note that $\triangle EBC$ is also an equilateral triangle, therefore $EC = BC = EB = x$. On the other hand, given that
$$x + 1 - x + 1 - x + 1 = 2\frac{1}{3}.$$
Hence, we deduce that
$$x = \frac{2}{3}.$$
Thus, it follows that
$$Area(ABCD) = 1^2 \cdot \frac{\sqrt{3}}{4} - x^2 \cdot \frac{\sqrt{3}}{4} = \frac{\sqrt{3}}{4}\left(1 - \frac{4}{9}\right) = \frac{5\sqrt{3}}{36}.$$

□

Problem 4.123. *Let point (1, 2) be the vertex of the graph of the first quadratic function and point (3, 6) be the vertex of the graph of the second quadratic function. The graph of the second function passes through the vertex of the graph of the first function and the graph of the first function passes through the vertex of the graph of the second function. What is the length of the line segment formed by two intersection points of the x–axis with the graph of the second function?*

(A) 6 (B) 2 (C) $2\sqrt{6}$ (D) $\sqrt{6}$ (E) $2\sqrt{5}$

Solution. Answer. (C)

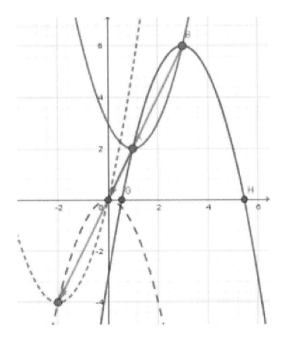

Let us translate both parabolas by the vector $\vec{v} = (-3, -6)$. In this case, the vertex of the second parabola translates to the point $(3 - 3, 6 - 6) = (0, 0)$ and the vertex of the first parabola to the point $(1 - 3, 2 - 6) = (-2, -4)$. Therefore, the 2nd parabola would pass through the points $(0, 0)$ and $(-2, -4)$. Thus, it follows that its graph is $y = -x^2$.

Hence, we obtain that the second parabola is described by the function $y = -(x - 3)^2 + 6$.

Therefore, the required length of the line segment is the length of $[x_1, x_2]$, where x_1 and x_2 are the roots of the equation $-(x-3)^2 + 6 = 0$. The roots are
$$x_1 = 3 - \sqrt{6}$$
and
$$x_2 = 3 + \sqrt{6}.$$
Thus, the required length of the line segment is
$$x_2 - x_1 = 2\sqrt{6}.$$

□

Problem 4.124. *Let n be a positive integer, denote by $A(n)$ be the number of digits of 2^n, by $B(n)$ be the number of digits of $2^{A(n)}$ and by $C(n)$ the number of digits of $2^{n+A(n)}$. How many elements are in the range of the function $A(n) + B(n) - C(n)$?*

(A) 1   (B) 5   (C) 2009   (D) 2010   (E) 2

Solution. Answer. (E)
Given that
$$\begin{cases} 10^{A(n)-1} < 2^n < 10^{A(n)}, \\ 10^{B(n)-1} < 2^{A(n)} < 10^{B(n)}, \\ 10^{C(n)-1} < 2^{n+A(n)} < 10^{C(n)}. \end{cases}$$
Multiplying the first two double inequalities, we obtain that
$$10^{A(n)+B(n)-2} < 2^{n+A(n)} < 10^{A(n)+B(n)}.$$
We have that
$$10^{C(n)-1} < 2^{n+A(n)} < 10^{C(n)}, 10^{A(n)+B(n)-2} < 2^{n+A(n)} < 10^{A(n)+B(n)}.$$
Thus, it follows that
$$C(n) - 1 < A(n) + B(n).$$
We deduce that
$$\begin{cases} C(n) \leq A(n) + B(n), \\ A(n) + B(n) - 2 < C(n). \end{cases}$$
Therefore
$$A(n) + B(n) - 1 \leq C(n).$$
Hence, it follows that
$$0 \leq A(n) + B(n) - C(n) \leq 1.$$
We have that
$$A(1) + B(1) - C(1) = 1$$
and
$$A(3) + B(3) - C(3) = 0.$$
Thus, the range of the function
$$A(n) + B(n) - C(n)$$
contains 2 elements.

□

Problem 4.125. Let (x_n) be a number sequence defined as follows: $x_1 = 1, x_2 = \frac{1}{2}, x_3 = \frac{1}{2}$ and $x_{n+3} = 2x_{n+2}x_{n+1} - x_n$. How many three-digit numbers n are there, such that $x_n = 1$?

(A) 0   (B) 75   (C) 900   (D) 100   (E) 76

Solution. Answer. (B)
Let us find some terms of the sequence (x_n). Note that

$$x_4 = 2 \cdot \frac{1}{2} \cdot \frac{1}{2} - 1 = -\frac{1}{2},$$

$$x_5 = 2 \cdot \frac{1}{2} \cdot \left(-\frac{1}{2}\right) - \frac{1}{2} = -1,$$

$$x_6 = \frac{1}{2},$$

$$x_7 = -\frac{1}{2},$$

$$x_8 = \frac{1}{2},$$

$$x_9 = -1,$$

$$x_{10} = -\frac{1}{2},$$

$$x_{11} = \frac{1}{2},$$

$$x_{12} = \frac{1}{2},$$

$$x_{13} = 1,$$

$$x_{14} = \frac{1}{2},$$

$$x_{15} = \frac{1}{2}.$$

Hence $x_{n+12} = x_n$, for $n = 1, 2, 3, ...$
Therefore, we have to find three-digit terms of the arithmetic sequence $1, 13, 25, ...$
The common term of this arithmetic sequence is

$$a_n = 1 + 12(n-1).$$

Thus, it follows that
$$100 \leq 1 + 12(n-1) \leq 999,$$

or

$$10 = \left\lceil \frac{111}{12} \right\rceil \leq n \leq \left\lfloor \frac{1010}{12} \right\rfloor = 84.$$

Therefore, there are 75 such three-digit numbers (as $84 - 10 + 1 = 75$). $\square$

4.6 Solutions of AMC 12 type practice test 6

Problem 4.126. *A train left city A at 8 : 00 AM and arrived in city B after 45 minutes. It stopped in city B for 10 minutes and continued on to city C. The train covered the distance between city B and city C in 75 minutes. At what time did the train arrive in city C?*

(A) 11:00 AM   (B) 10:05 AM   (C) 10:00 AM   (D) 10:10 AM   (E) 10:30 AM

Solution. Answer. (D)
The train covered the distance between cities A and C in $45 + 10 + 75 = 130$ minutes or 2 hours and 10 minutes. Thus, it follows that the train reached city C at 10:10 AM. $\square$

Problem 4.127. *What is the opposite of the value of the following expression $\frac{1}{3} - \frac{3}{4}$?*

(A) -2.4 (B) $-\frac{5}{12}$ (C) $\frac{5}{12}$ (D) 1 (E) $\frac{1}{2}$

Solution. Answer. (C)
We have that
$$\frac{1}{3} - \frac{3}{4} = -\frac{5}{12}.$$
The opposite of this number is $\frac{5}{12}$. $\square$

Problem 4.128. *Six painters working at the same constant rate can completely paint an apartment in 8 hours. How many painters were working if it took 12 hours to paint the apartment?*

(A) 5   (B) 3   (C) 4   (D) 2   (E) 1

Solution. Answer. (C)
Given that six painters are painting (at the same rate) an apartment in 8 hours, therefore one painter can paint the same apartment in 48 hours. Thus, it follows that four painters can paint the apartment in 12 hours. $\square$

Problem 4.129. *What is the value of the following expression?*
$$\frac{1^2 + 1 \cdot 2 + 2^2}{1^3 \cdot 2^3} + \frac{2^2 + 2 \cdot 3 + 3^2}{2^3 \cdot 3^3} + \ldots + \frac{9^2 + 9 \cdot 10 + 10^2}{9^3 \cdot 10^3}.$$

(A) 1   (B) 0.9   (C) 0.99   (D) 0.5   (E) 0.999

Solution. Answer. (E)
Note that
$$\frac{k^2 + k \cdot (k+1) + (k+1)^2}{k^3 \cdot (k+1)^3} = \frac{(k+1-k)(k^2 + k \cdot (k+1) + (k+1)^2)}{k^3 \cdot (k+1)^3} =$$
$$= \frac{(k+1)^3 - k^3}{k^3 \cdot (k+1)^3} = \frac{1}{k^3} - \frac{1}{(k+1)^3},$$
for $k = 1, 2, \ldots, 9$.
Therefore, the required sum is
$$1 - \frac{1}{2^3} + \frac{1}{2^3} - \frac{1}{3^3} + \ldots + \frac{1}{9^3} - \frac{1}{10^3} = 1 - \frac{1}{10^3} = 0.999.$$

$\square$

Problem 4.130. *For how many positive values of x is $\dfrac{20}{x+1} + \dfrac{1}{3(x+1)}$ a positive integer?*

(A) 0   (B) 10   (C) 19   (D) 20   (E) 25

Solution. Answer. (D)
Note that
$$\frac{20}{x+1} + \frac{1}{3(x+1)} = \frac{61}{3(x+1)} < \frac{61}{3}.$$
Therefore, the fraction
$$\frac{61}{3(x+1)}$$
can attain natural values from 1 to 20.
Let $n \in \{1, 2, ..., 20\}$ and
$$\frac{61}{3(x+1)} = n.$$
Hence we obtain that
$$x = \frac{61 - 3n}{3n} > 0.$$
$\square$

Problem 4.131. *A shop bought a coat for $500. The shop inteded to sell the coat for some price, but sold it 5% more than the intended sales price. Given that for that deal the shop generated a total profit of $67. What was the intended sales price of the coat?*

(A) $560   (B) $540   (C) $550   (D) $530   (E) $520

Solution. Answer. (B)
Let $\$x$ be the intended sales price of the coat. Thus, it follows that
$$1.05x - 67 = 500.$$
Therefore $x = 540$. $\square$

Problem 4.132. *A fisherman makes a round-trip and takes his boat 10 miles into a lake from its shore and back (covering the same distance). The average speed of the boat is 2.5 miles per hour. The fisherman catches (in average) 0.5 pounds of fish per hour. How many pounds of fish does the fisherman catch per one round-trip?*

(A) 4   (B) 3   (C) 2   (D) 5   (E) 6

Solution. Answer. (A)
Note that the fisherman boats in the lake $\dfrac{20}{2.5} = 8$ hours. So, one-round trip takes 8 hours.
Given that the fisherman catches (in average) 0.5 pounds of fish per hour.
Therefore, the fisherman catches $8 \cdot 0.5 = 4$ pounds of fish per one round-trip. $\square$

Problem 4.133. *Given a rectangular prism, such that its volume is equal to $\sqrt{3}$. Each edge of the rectangular prism is increased $\sqrt{3}$ times. What is the volume of the new rectangular prism?*

(A) $3\sqrt{3}$ (B) 9   (C) 3   (D) 27   (E) $9\sqrt{3}$

Solution. Answer. (B)
Let the dimensions of the initial rectangular prism be a, b, c. Given that $abc = \sqrt{3}$. The dimensions of the new rectangular prism will be $a\sqrt{3}, b\sqrt{3}, c\sqrt{3}$ and the volume is
$$a\sqrt{3} \cdot b\sqrt{3} \cdot c\sqrt{3} = 9.$$

□

Problem 4.134. *The width and length of the frame of a painting are in the proportion 4:5. The respective dimensions of the painting are in the proportion 3:4 and the painting is the same distance from the frame on each side. What is the ratio of the area of the frame to the area of the painting?*

(A) $\dfrac{1}{2}$ (B) $\dfrac{2}{3}$ (C) $\dfrac{1}{3}$ (D) $\dfrac{1}{4}$ (E) $\dfrac{1}{10}$

Solution. Answer. (B)

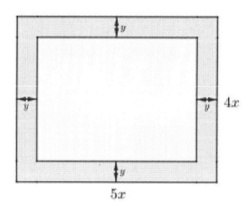

Let the width of the frame be $4x$ and the length be $5x$. Given that
$$\frac{5x - 2y}{4x - 2y} = \frac{4}{3}.$$

Thus, it follows that $x = 2y$. Therefore, the ratio of the area of the frame to the area of the painting is:
$$\frac{20x^2 - (5x - 2y)(4x - 2y)}{(5x - 2y)(4x - 2y)} = \frac{80y^2 - 48y^2}{48y^2} = \frac{2}{3}.$$

Alternative solution. Let the width of the painting be $3z$ and the length be $4z$. Thus, it follows that
$$5x - 4z = 4x - 3z.$$

Hence, we obtain that $x = z$. Therefore, we can take $x = z = 1$. Thus, the area of the painting is 12sq. units and the area of the frame is $5 \cdot 4 - 3 \cdot 4 = 8$ sq. units. Hence, the ratio of the area of the frame to the area of the painting is:
$$\frac{8}{12} = \frac{2}{3}.$$
□

Problem 4.135. *One combine harvester can harvest a field on its own in 12 hours. A second combine harvester can harvest the same field on its own in 18 hours. How many hours will it take both combine harvesters, working together, to harvest the whole field, if they also take a 48 minute break?*

(A) 7.2   (B) 7   (C) 9   (D) 8   (E) 10

Solution. Answer. (D)
First combine harvester's rate is $\frac{1}{12}$ of the field per hour, while the same for the second combine harvester is $\frac{1}{18}$ per hour. Therefore, both can harvest
$$\frac{1}{12} + \frac{1}{18} = \frac{5}{36}$$
of the field per hour.
Therefore, it would take them $\frac{36}{5} = 7.2$ hours to complete the job and if they take rest for 48 min (0.8 hours), then they can complete the job in $7.2 + 0.8 = 8$ hours. □

Problem 4.136. *All faces of four identical $1 \times 1 \times 1$ cubes are numbered from 1 to 6. Moreover, the opposite faces are numbered with numbers 1 and 2, 3 and 4, 5 and 6. These four cubes are used to form a $1 \times 1 \times 4$ rectangular prism (placed on the table), such that the sum of 17 numbers on the visible faces is the smallest possible. What is the value of the sum of these 17 numbers?*

(A) 44   (B) 40   (C) 45   (D) 42   (E) 74

Solution. Answer. (C)
Note that the sum of all the numbers on all faces is $4 \cdot 21 = 84$. There are 7 numbers that are not visible. Moreover, the numbers on two opposite faces of three cubes are not visible, and only one of the cubes has a face with the number that is not visible.
Therefore, the greatest possible sum of the numbers that are not visible is
$$3 \cdot (5 + 6) + 6 = 39.$$
Note that the sum of the visible numbers is the least possible if the sum of non-visible numbers is the greatest possible. Thus, the required number is $84 - 39 = 45$. □

Problem 4.137. *Let function $f(x)$ be defined on $[0, 1]$. Given that the sum of the greatest and the smallest values is equal to 10. What is the value of the sum of the greatest and the smallest values of function $g(x) = 13 - 2 \cdot f(1 - x)$?*

(A) -7 (B) 7   (C) 26   (D) 13   (E) 6

Solution. Answer. (E)
Let the greatest and the least values of $f(x)$ be M and m, respectively. Moreover, assume that
$$f(x_0) = M, f(x_1) = m$$
and
$$x_0, x_1 \in [0, 1].$$

Given that $M + m = 10$. For any $x \in [0,1]$ we have that
$$m \le f(x) \le M.$$
Hence, for any $x \in [0,1]$, we obtain that
$$m \le f(1-x) \le M.$$
From the last inequality we deduce that
$$-2M \le -2f(1-x) \le -2m.$$
Thus, it follows that
$$13 - 2M \le 13 - 2f(1-x) \le 13 - 2m.$$
Hence, for any $x \in [0,1]$, we have that
$$13 - 2M \le g(x) \le 13 - 2m.$$

Note that
$$g(1 - x_0) = 13 - 2f(x_0) = 13 - 2M$$
and
$$g(1 - x_1) = 13 - 2f(x_1) = 13 - 2m.$$

Therefore, the greatest and the least values of $g(x)$ are $13 - 2m$ and $12 - 2M$, respectively. Thus, it follows that
$$13 - 2m + 13 - 2M = 26 - 2(m+M) = 26 - 2 \cdot 10 = 6.$$

$\square$

Problem 4.138. *Let AB be the diameter of a semicircle with center O. Semicircles, with diameters OA and OB, are drawn. Circle S is drawn, such that it is tangent to these three semicircles (see the figure). Given that the radius of circle S is equal to r. What is the value of $\dfrac{AB}{r}$?*

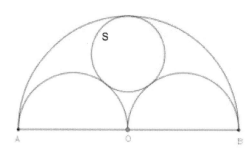

(A) 5   (B) 4   (C) 6   (D) 4.5   (E) 3

Solution. Answer. (C)
Let us denote the centers of semicircles with diameters OA and OB by O_1 and O_2, respectively (see the figure). Let us denote the center of circle S by O_3 and let $OA = R$.

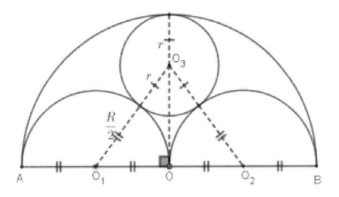

Thus, it follows that
$$O_2O_3 = O_1O_3 = \frac{R}{2} + r,$$
$$OO_3 = R - r,$$

and
$$OO_1 = OO_2 = \frac{R}{2}.$$

Hence, we obtain that $OO_3 \perp O_1O_2$.
Therefore, from the right triangle OO_1O_3 according to the Pythagorean theorem, we obtain that
$$\left(\frac{R}{2} + r\right)^2 = \left(\frac{R}{2}\right)^2 + (R - r)^2.$$

We deduce that $R = 3r$ or
$$\frac{AB}{r} = \frac{2R}{r} = 6.$$

□

Problem 4.139. *Let x, y be integers. What is the total number of all points $M(x, y)$, such that the following inequality holds true?*
$$|4x - 24| + |3y + 15| \leq 12.$$

(A) 27 (B) 32 (C) 25 (D) 29 (E) 28

Solution. Answer. (A)
$x - 6 = x'$ and $y + 5 = y'$, in this case we should find the number of all points $M'(x', y')$, such that
$$4|x'| + 3|y'| \leq 12$$

and x', y' are integers.
If $|x'| = 0$, then the number of points M' is 9.
If $|x'| = 1$, then the number of points M' is $2 \cdot 5$.
If $|x'| = 2$, then the number of points M' is $2 \cdot 3$.
If $|x'| = 3$, then the number of points M' is $2 \cdot 1$.
Therefore, the total number of all points $M(x, y)$, such that given inequality holds true is equal to 27. □

Problem 4.140. *What is the units digit of the following number?*

$$2008^{2007^{2008}} + 2007^{2008^{2007}}.$$

(A) 1   (B) 3   (C) 5   (D) 7   (E) 9

Solution. Answer. (E)
Note that
$$4 \mid 2007^{2008} - 1 = (2007^2)^{1004} - 1^{1004}.$$

$2008^{2007^{2008}} = 2008^{4k+1}$, where k is a positive integer. Note that as 2008^4 ends with the digit 6, then 2008^{4k} also ends with the digit 6. Therefore $2008^{2007^{2008}}$ ends with the digit 8.
We have that $4 \mid 2008^{2007}$, hence

$$2007^{2008^{2007}} = 2007^{4m} = (2007^4)^m,$$

where m is a positive integer.
Note that as 2007^4 ends with the digit 1, then $2007^{2008^{2007}}$ also ends with the digit 1.
Therefore $2008^{2007^{2008}} + 2007^{2008^{2007}}$ ends with the digit 9.   □

Problem 4.141. *Let a, b be numbers greater than 1, such that the following three numbers*

$$\log(a^2 b), \log(a^2 b^2), \log(a^5 b^2)$$

form a geometric sequence. What is the value of the common ratio of this geometric sequence?

(A) $\dfrac{2}{3}$ (B) 1.5   (C) 2   (D) 0.5   (E) 3

Solution. Answer. (B)
Let $\log a = x, \log b = y$. Given that $x > 0, y > 0$ and the numbers $2x + y, 2x + 2y, 5x + 2y$ form a geometric sequence. Thus, it follows that
$$(2x + 2y)^2 = (2x + y)(5x + 2y).$$

Hence, we deduce that
$$6\left(\frac{x}{y}\right)^2 + \frac{x}{y} - 2 = 0.$$

We obtain that
$$\frac{x}{y} = \frac{1}{2}.$$

Therefore, the answer is
$$\frac{2x + 2y}{2x + y} = \frac{6x}{4x} = 1.5.$$
    □

Problem 4.142. *Let n be a positive integer and (a_n) be a sequence defined as follows: $a_1 = 2430$ and*

$$a_n = \begin{cases} \dfrac{a_{n-1}}{3}, & \text{if } 3 \mid a_{n-1}, \\ 2a_{n-1} + 1, & \text{if } 3 \nmid a_{n-1}, \end{cases}$$

for $n = 2, 3, \ldots$. How many numbers among the terms $a_1, a_2, \ldots, a_{2008}$ are divisible by 3?

(A) 662   (B) 104   (C) 6   (D) 9   (E) 7

Solution. Answer. (E)
We have that
$$a_1 = 2430, a_2 = 810, a_3 = 270, a_4 = 90, a_5 = 30, a_6 = 10, a_7 = 21, a_8 = 7, a_9 = 15, a_{10} = 5, \ldots$$

Note that each of the terms $a_{10}, a_{11}, \ldots, a_{2008}$ leaves a remainer of 2 after division by 3.
Thus, it follows that
$$a_{10} = 3q + 2.$$

Hence, we deduce that
$$a_{11} = 2a_{10} + 1 = 2(3q + 2) + 1 = 3(2q + 1) + 2.$$

In a similar way, we obtain that each of the numbers $a_{12}, a_{13}, \ldots, a_{2008}$ leaves a remainder of 2 after division by 3.
Therefore, the answer is 7. □

Problem 4.143. *Let $SABC$ be a pyramid, such that $AB = AC = 13, BC = 5$ and $\angle ASB = \angle ASC = 90°, \angle BSC = 60°$. What is the value of the volume of $SABC$?*

(A) 50   (B) 25   (C) $25\sqrt{3}$ (D) $75\sqrt{3}$ (E) $50\sqrt{3}$

Solution. Answer. (C)
Note that $\triangle ASB$ and $\triangle ASC$ are congruent triagles, therefore $SB = SC$. Taking this into consideration and from the given condition ($\angle BSC = 60°$) we obtain that $\triangle SBC$ is an equilateral triangle, therefore $SB = SC = 5$.
From triangle ASB according to the Pythagorean theorem, we obtain that
$$SA^2 + SB^2 = AB^2.$$

Thus, it follows that $SA = 12$.
Note that $SA \perp SB$ and $SA \perp SC$. Hence, we obtain that SA is perpendicular to plane SBC. Therefore, the volume of the pyramid $SABC$ is
$$\frac{1}{3} \cdot 12 \cdot \frac{1}{2} \cdot 5 \cdot 5 \cdot \sin 60° = 25\sqrt{3}.$$
□

Problem 4.144. *What is the coefficient of x^{23} in the expansion of the following expression?*
$$(24 + 23 \cdot x + 22 \cdot x^2 + \ldots + 1 \cdot x^{23})(1 + x + x^2 + \ldots + x^9)(1 + x + x^2 + \ldots + x^{11}).$$

(A) 1320   (B) 132   (C) 529   (D) 1000   (E) 5290

Solution. Answer. (A)
Let us consider the following polynomial
$$p(x) = (1 + x + \ldots + x^9)(1 + x + \ldots + x^{11}) = a_0 + a_1 x + \ldots + a_{20} x^{20}.$$

Therefore, the given product is equal to the following product
$$(24 + 23 \cdot x + 22 \cdot x^2 + \ldots + x^{23})(a_0 + a_1 x + \ldots + a_{20} x^{20}).$$

Note that the coefficient of x^{23} is $a_0 + 2a_1 + \ldots + 21 a_{20}$.
We have that
$$xp(x) = a_0 x + a_1 x^2 + \ldots + a_{20} x^{21}.$$

Thus, it follows that
$$(xp(x))' = a_0 + 2a_1 x + \ldots + 21 a_{20} x^{20}.$$

On the other hand, we have that

$$(xp(x))' = ((x+...+x^{10})(1+x+...+x^{11}))' = (1+...+10x^9)(1+x+...+x^{11})+(x+...+x^{10})(1+...+11x^{10}).$$

From the last two equations, we deduce that

$$a_0 + 2a_1x + ... + 21a_{20}x^{20} = (1 + ... + 10x^9)(1 + x + ... + x^{11}) + (x + ... + x^{10})(1 + ... + 11x^{10}).$$

If $x = 1$, then obtain that

$$a_0 + 2a_1 + ... + 21a_{20} = 55 \cdot 12 + 10 \cdot 66 = 1320.$$

□

Problem 4.145. *Let CM be a median in a right triangle with legs AC and BC. Given that $AC = 6$ and $BC = 8$. What is the value of the sum of the inradii of triangles ACM and BCM?*

(A) $2\frac{1}{3}$ (B) 3.5 (C) $2\frac{5}{6}$ (D) 3 (E) 2

Solution. Answer. (C)
We have that
$$Area(ACM) = Area(BCM) = \frac{Area(ABC)}{2} = \frac{1}{2} \cdot \frac{6 \cdot 8}{2} = 12.$$

Moreover, according to the Pythagorean theorem we obtain that

$$AB^2 = AC^2 + BC^2 = 100.$$

Hence, we deduce that $AB = 10$.
We have that
$$CM = \frac{AB}{2} = 5.$$

The inradii of triangles ACM and BCM are equal to:

$$\frac{12}{\frac{6+5+5}{2}}$$

and

$$\frac{12}{\frac{8+5+5}{2}},$$

respectively. Thus, their sum is:

$$\frac{12}{8} + \frac{12}{9} = 2\frac{5}{6}.$$

□

Problem 4.146. *How many five-digit numbers are there which are divisible by 11 and are composed (without repetitions) only of digits 1,2,3,4,8?*

(A) 10 (B) 12 (C) 24 (D) 11 (E) 0

Solution. Answer. (B)
Let $\overline{abcde}$ five-digit number be divisible by 11. We have that

$$\overline{abcde} = 9999a + 1001b + 99c + 11d + a - b + c - d + e.$$

Therefore, $\overline{abcde}$ is divisible by 11 if and only if
$$a - b + c - d + e = -11,$$
or
$$a - b + c - d + e = 0,$$
or
$$a - b + c - d + e = 11.$$
The first and the last cases are impossible, as $a - b + c - d + e$ and $a + b + c + d + e = 18$ have the same parity. Hence, we obtain that
$$a + c + e = 9,$$
and
$$b + d = 9.$$
Thus, it follows that $\{a, c, e\} = \{2, 3, 4\}$ and $\{b, d\} = \{1, 8\}$. Therefore, the total number of five-digit numbers satisfying the assumptions of the problem is $3! \cdot 2 = 12$. $\square$

Problem 4.147. *Let a square be constructed externally on each side of a regular hexagon with a side length of 1. What is the radius of the circle which passes through all the vertices of these squares that are not the vertices of the hexagon?*

(A) $\sqrt{3}$ (B) 2   (C) $\dfrac{\sqrt{6} + \sqrt{2}}{2}$ (D) $\sqrt{5} + 1$ (E) $\sqrt{2}$

Solution. Answer. (C)

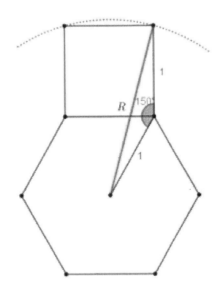

From the law of cosines (see the figure) we have that
$$R^2 = 1^2 + 1^2 - 2 \cdot 1 \cdot 1 \cdot \cos 150° = 2 + \sqrt{3}.$$
Thus, it follows that
$$R = \sqrt{2 + \sqrt{3}} = \sqrt{\frac{4 + 2\sqrt{3}}{2}} = \frac{\sqrt{3} + 1}{\sqrt{2}} = \frac{\sqrt{6} + \sqrt{2}}{2}.$$
$\square$

Problem 4.148. *Given that the solutions of the following equation are the vertices of a quadrilateral on the complex plane.*
$$(z+i)^4 + (z-i)^4 = 16.$$
What is the value of the perimeter of that quadrilateral?

(A) 20   (B) 8   (C) $2\sqrt{5}$ (D) $8\sqrt{5}$ (E) $8\sqrt{2}$

Solution. Answer. (E)
We have that
$$(z+i)^4 + (z-i)^4 = ((z+i)^2 + (z-i)^2)^2 - 2(z+i)^2(z-i)^2 =$$
$$= (2z^2 - 2)^2 - 2(z^2+1)^2 = 2z^4 - 12z^2 + 2.$$

Taking this into consideration and from given equation, we obtain the following equation.
$$z^4 - 6z^2 - 7 = 0.$$

Note that the roots of this equation are $\sqrt{7}, -\sqrt{7}, i, -i$.
Hence, the perimeter of the quadrilateral with the mentioned vertices is equal to $8\sqrt{2}$. □

Problem 4.149. *Let D be a given point on side BC of triangle ABC. Given that $CD = 1, BD = 3$ and $\angle C = \dfrac{\pi}{3}$. What is the greatest possible value of the measure of $\angle BAD$?*

(A) $\dfrac{3\pi}{8}$ (B) $\dfrac{\pi}{6}$ (C) $\dfrac{\pi}{2}$ (D) $\dfrac{\pi}{3}$ (E) $\dfrac{\pi}{4}$

Solution. Answer. (D)
Let us draw the altitude BH of triangle ABC and the altitude EH of triangle CBH (see the figure).

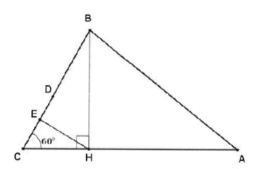

As $\angle CBH = 30°$, then from right triangle CBH we obtain that
$$CH = \frac{BC}{2} = 2.$$

Note that in right triangle CHE we have that $\angle CHE = 30°$, therefore
$$CE = \frac{CH}{2} = 1 = CD.$$

Hence, we obtain that points D and E coincide (see the figure).

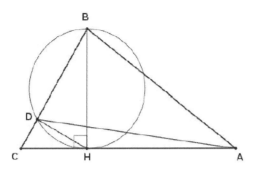

As $\angle BDH = 90°$, thus BH is the diameter of the circumcircle of triangle BDH. Hence, the circumcenter of triangle BDH is the midpoint of line segment BH.

Therefore, that circle is tangent to AC at point H. We deduce that

$$\angle BAD \leq \frac{minorarc(BD)}{2} = \angle BHD = \frac{\pi}{3}.$$

Hence, the greatest possible measure of $\angle BAD$ is equal to $\frac{\pi}{3}$. Note that $\frac{\pi}{3}$ is a reachable value for $\angle BAD$, as one can provide an example of triangle ABC, such that $CD = 1, BD = 3, \angle C = \frac{\pi}{3}, \angle BAD = \frac{\pi}{3}$. □

Problem 4.150. *Let n be a positive integer and $(a_n), (b_n)$ be sequences of complex numbers defined as follows: $a_1 = i, b_1 = 1$ and*

$$a_{n+1} = \sqrt[3]{2}a_n - b_n,$$
$$b_{n+1} = a_n + \sqrt[3]{2}b_n,$$

where $n = 1, 2, ...$. What is the value of the expression $\left|\frac{a_{2008}}{b_{2008}}\right|$?

(A) $\sqrt[3]{2}$ (B) 1   (C) 2   (D) $\frac{1}{2}$ (E) $\sqrt[3]{4}$

Solution. Answer. (B)

Note that

$$(a_{n+1})^2 + (b_{n+1})^2 = (\sqrt[3]{4} + 1)(a_n^2 + b_n^2), n = 1, 2, ...$$

Hence, applying this formula recursively, we obtain that

$$(a_{n+1})^2 + (b_{n+1})^2 = (\sqrt[3]{4} + 1)^n(a_1^2 + b_1^2), n = 1, 2, ...$$

Plugging in $n = 2007$ in the last equation and using that $a_1^2 + b_1^2 = 0$ (as $a_1 = i, b_1 = 1$), we deduce that

$$(a_{2008})^2 + (b_{2008})^2 = 0.$$

Note that $b_{2008} \neq 0$, as otherwise if $b_{2008} = 0$, then from the last equation we deduce that

$$b_{2008} = a_{2008} = 0.$$

In this case, we obtain that

$$b_{2007} = a_{2007} = 0, ..., b_1 = a_1 = 0.$$

This leads to a contradiction, as $a_1 = i, b_1 = 1$. Therefore, we deduce that $b_{2008} \neq 0$ and

$$(a_{2008})^2 = -(b_{2008})^2.$$

Thus, it follows that

$$\left|\frac{a_{2008}}{b_{2008}}\right| = 1.$$

□

4.7 Solutions of AMC 12 type practice test 7

Problem 4.151. *The area of a park is 300 sq.m and the area of a nearby park is 430 sq.m. The area of the first park is p% less than the area of the second park. What is the value of the closest integer to p?*

(A) 29   (B) 30   (C) 31   (D) 32   (E) 33

Solution. Answer. (B)

The first land is 130 sq.m. less than the area of the second land, which is $\frac{130}{430}$ part of the latter land. Therefore
$$p = \frac{13}{43} \cdot 100 = \frac{1300}{43} = 30\frac{10}{43},$$
and the answer is 30, as
$$\frac{10}{43} < \frac{1}{2}.$$

□

Problem 4.152. *Let positive integer a be 25% less than a positive integer b. Given that their sum is equal to 49. What is the value of a?*

(A) 21   (B) 30   (C) 27   (D) 14   (E) 35

Solution. Answer. (A)
Given that
$$a = \frac{75}{100}b = \frac{3}{4}b,$$
and
$$a + b = 49.$$
Thus, it follows that
$$a + \frac{4}{3}a = 49.$$
Therefore $a = 21$.

□

Problem 4.153. *What is the smallest possible perimeter of a triangle with different integer side lengths?*

(A) 5   (B) 6   (C) 12   (D) 8   (E) 9

Solution. Answer. (E)
Note that the largest side of a triangle with different integer side lengths is not less than 4, as a triangle with side lengths 1, 2, 3 does not exist. Therefore, the sum of the lengths of two other sides is not less than 5. Thus, the smallest possible perimeter is 9.
For example, the perimeter of a triangle with side lengths 2, 3, 4 is equal to 9.

□

Problem 4.154. *How many positive odd divisors does 30^{10} have?*

(A) 121   (B) 1331   (C) 665   (D) 666   (E) 667

Solution. Answer. (A)
We have that
$$30^{10} = 2^{10} \cdot 3^{10} \cdot 5^{10}.$$
Thus, it follows that any odd divisor of this number is of the form $3^a \cdot 5^b$, where $a, b \in \{0, 1, 2, ..., 10\}$. Therefore, there are $11 \cdot 11 = 121$ odd divisors.

□

Problem 4.155. *Given two congruent circles on the plane with non-coincident centers. How many of the transformations below map one circle onto another?*
- *parallel translation.*
- *point symmetry.*
- *line symmetry.*
- *rotation by 30° angle.*

(A) 4   (B) 3   (C) 2   (D) 1   (E) 0

Solution. Answer. (A)
All 4 possible cases are drawn in the figure below.

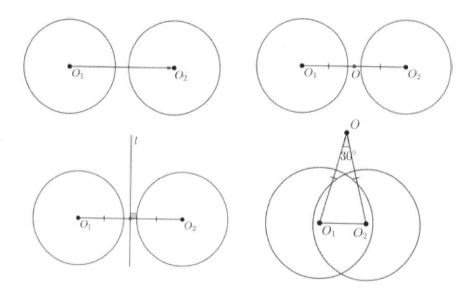

Therefore, 4 of the transformations given in the problem map one circle onto another. $\square$

Problem 4.156. *What is the value of the greatest integer that cannot be written as a sum of two composite numbers?*

(A) 101   (B) 11   (C) 111   (D) 2019   (E) 42

Solution. Answer. (B)
If n is an even number and $n \geq 8$. We have that $n = 4 + (n-4)$, where 4 and $n-4$ are composite numbers.
If n is an odd number and $n \geq 13$. We have that $n = 9 + (n-9)$, where 9 and $n-9$ are composite numbers.
Note that 11 cannot be written as a sum of two composite numbers and it is the largest with this property. $\square$

Problem 4.157. Let ABC be a triangle, such that $AC = 10, BC = 17$ and $AB = 21$. Let M and N be points on side AB, such that $CM^2 = AM \cdot BM$ and $CN^2 = AN \cdot BN$ (see the figure). What is the value of the area of triangle CMN?

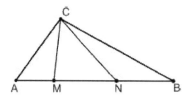

(A) 34 (B) 42 (C) 30 (D) 36 (E) $10\sqrt{3}$

Solution. Answer. (A)

Let us choose standard rectangular coordinate system (as shown in the figure).

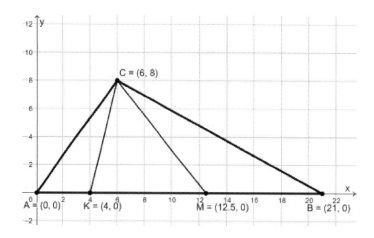

Let $C(x_0, y_0)$. Given that
$$x_0^2 + y_0^2 = 100,$$
and
$$(x_0 - 21)^2 + y_0^2 = 289.$$
Thus, it follows that
$$42x_0 - 441 = 100 - 289,$$
or $x_0 = 6$ and $y_0 = 8$. Therefore $K(x, 0)$ lies on segment AB, such that
$$CK^2 = AK \cdot BK.$$
Hence
$$(x - 6)^2 + 8^2 = x(21 - x).$$
We obtain that $x = 4$ or $x = 12.5$. Thus, it follows that $MN = 12.5 - 4 = 8.5$ and
$$Area(CMN) = \frac{8 \cdot 8.5}{2} = 34.$$

Alternative solution. This problem can also be solved using Stewart's theorem. An interested reader can do it independently. □

Problem 4.158. *A geometric sequence consists of five terms. The arithmetic mean of the first four terms of this sequence is equal to 10. The arithmetic mean of the last four terms is equal to 30. What is the fifth term of this sequence?*

(A) 40 (B) 52 (C) 64 (D) 72 (E) 81

Solution. Answer. (E)
Let the terms of the geometric sequence be b_1, b_2, b_3, b_4 and b_5. Given that
$$\frac{b_1 + b_2 + b_3 + b_4}{4} = 10,$$
and
$$\frac{b_2 + b_3 + b_4 + b_5}{4} = 30.$$
We have that
$$b_2 + b_3 + b_4 + b_4 = (b_1 + b_2 + b_3 + b_4) \cdot r,$$
where r is the common ratio of the geometric sequence. Therefore, $10 \cdot r = 30$ or $r = 3$. Hence, we obtain that
$$b_1 + 3b_1 + 9b_1 + 27b_1 = 40.$$
Therefore $b_1 = 1$. Thus, it follows that
$$b_5 = b_1 \cdot r^4 = 81.$$

□

Problem 4.159. *Consider four congruent circles on a plane, such that they are pairwise non-concentric and they divide the plane in n parts. How many values of n are possible?*

(A) 10 (B) 9 (C) 8 (D) 7 (E) 6

Solution. Answer. (A)
Note that m congruent circles divide a plane into at least $m + 1$ parts as each new circle increases the part by at least 1. On the other hand, these m circles divide the plane into at most $m^2 - m + 2$ parts at m^{th} circle can increase the number of parts by not more than $2m - 2$.

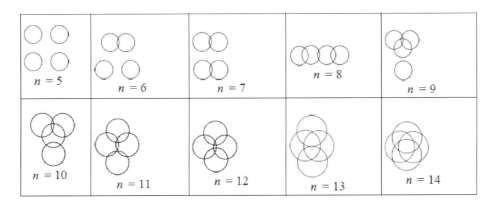

□

Problem 4.160. *Let x and y be randomly chosen numbers from $[0,1]$. What is the probability that the following inequality holds true?*
$$|x - 0.5| + |y - 1.5| \leq 1.$$

(A) $\dfrac{1}{4}$ (B) $\dfrac{3}{4}$ (C) $\dfrac{1}{6}$ (D) $\dfrac{5}{6}$ (E) $\dfrac{1}{2}$

Solution. Answer. (A)

Let (x,y) be a point in the square $OABC$ ($[0,1] \times [0,1]$) in the coordinate plane. We have that $y \in [0,1]$.

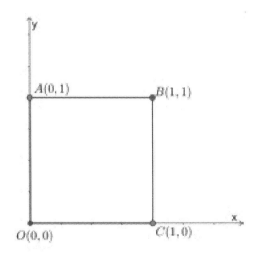

Thus, it follows that $y - 1.5 < 0$ and the given inequality is equivalent to the following inequality:
$$|x - 0.5| \leq y - 0.5,$$
or equivalently
$$0.5 - y \leq x - 0.5 \leq y - 0.5.$$

Hence, we obtain that $y \geq x$ and $y \geq 1 - x$. The solution of the last two inequalities is the shaded region ($\triangle ABD$).

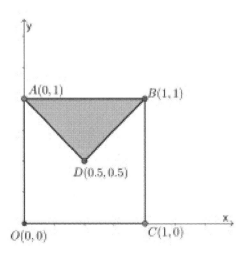

Therefore, the required probability is equal to $\dfrac{1}{4}$. □

Problem 4.161. *Given 22 circles, such that 21 of them have radius 1 (see the figure). Given also that any two circles that have a common point are pairwise tangent. What is the value of the circumference of the largest circle?*

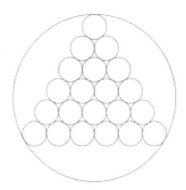

(A) $\dfrac{2\pi}{3}(10\sqrt{3}+3)$ (B) $\dfrac{20}{3}\pi$ (C) $\dfrac{20\sqrt{3}}{3}\pi$ (D) 13π (E) 20π

Solution. Answer. (A)
We have equilateral triangles with side length 2 (see the figure).

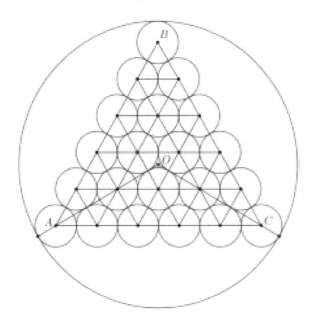

Therefore
$$AB = BC = AC = 10.$$
Let O be the center of the largest circle and R its radius. Then O is $(R-1)$ away from points A, B and C.
Thus, it follows that
$$R - 1 = \dfrac{10}{\sqrt{3}},$$
and
$$2\pi R = \dfrac{2\pi}{3}(10\sqrt{3}+3).$$

Problem 4.162. *Let x and y be positive numbers, such that $x \neq 1$ and $y \neq 1$. Given that $\log_4 x = \log_y 3$ and $\log_3 y = \log_2 \frac{32}{x}$. What is the value of the expression $\log_x 2 + \log_y 3$?*

(A) 2   (B) 3   (C) 2.5   (D) 3.5   (E) 4

Solution. Answer. (C)
We have that
$$\log_4 x = \log_y 3.$$
Thus, it follows that
$$\frac{1}{2} \log_2 x = \frac{1}{\log_3 y},$$
and
$$\log_3 y = \log_2 \frac{32}{x}.$$
Hence, we obtain that
$$\log_3 y = 5 - \log_2 x.$$
We deduce that
$$\log_2 x \cdot \log_3 y = 2,$$
and
$$\log_2 x + \log_3 y = 5.$$
Therefore
$$\log_x 2 + \log_y 3 = \frac{1}{\log_2 x} + \frac{1}{\log_3 y} = \frac{\log_2 + \log_3}{\log_2 x \cdot \log_3 y} = \frac{5}{2}.$$

□

Problem 4.163. *Let a and b be complex numbers, such that for all values of x it holds true:*
$$(x^2 - 2x + 3)(x^2 - 4x + 12) = (x^2 + ax + 6)(x^2 + bx + 6).$$

What is the value of the expression $|a - b|$?

(A) $2\sqrt{2}$ (B) $3\sqrt{2}$ (C) $6\sqrt{2}$ (D) 12   (E) 2

Solution. Answer. (A)
Given that
$$x^4 - 6x^3 + 23x^2 - 36x + 36 = x^4 + (a+b)x^3 + (12+ab)x^2 + (6a+6b)x + 36.$$

Thus, it follows that $a + b = -6$ and $ab = 11$.
We have that
$$(a-b)^2 = (a+b)^2 - 4ab = -8.$$
Hence, we obtain that
$$|a - b| = 2\sqrt{2}.$$

□

Problem 4.164. Let n be a positive integer. Consider a number sequence (a_n), such that $a_1 = 1, a_2 = 5$ and
$$a_{n+2} = 5a_{n+1} - 6a_n.$$
where $n = 1, 2, \ldots$. What is the value of the expression $(a_{100} - 3 \cdot a_{99})$?

(A) 2^{99} (B) 2^{100} (C) 3^{99} (D) 3^{100} (E) $3^{100} + 2^{99}$

Solution. Answer. (A)
We have that
$$a_{n+2} - 3a_{n+1} = 2(a_{n+1} - 3a_n).$$
Let us define a sequence (b_n) in the following way
$$b_n = a_{n+1} - 3a_n.$$
Note that (b_n) is a geometric sequence with a common ratio 2. Therefore
$$a_{100} - 3a_{99} = b_{99} = 2^{98} \cdot b_1 = 2^{98}(5 - 3 \cdot 1) = 2^{99}.$$

Alternative solution 1. Consider the first terms of the given sequence for $n = 1, 2, \ldots$, then we can guess the following recursive rule $a_n = 3^n - 2^n$. It can be proven that such sequence satisfies the assumptions of the problem.

Alternative solution 2. The characteristic equation of the recursive formula of a_n is the following one:
$$x^2 - 5x + 6 = 0.$$
The roots of this equation are $x_1 = 2$ and $x_2 = 3$. Thus, it follows that
$$a_n = \alpha \cdot 2^n + \beta \cdot 3^n.$$
Now, from the initional conditions $a_1 = 1$ and $a_2 = 5$, we obtain that
$$\begin{cases} 2\alpha + 3\beta = 1, \\ 4\alpha + 9\beta = 5. \end{cases}$$
Therefore, we deduce that $\alpha = -1, \beta = 1$ and
$$a_n = 3^n - 2^n.$$
Thus, it follows that
$$a_{100} - 3 \cdot a_{99} = 3^{100} - 2^{100} - 3^{100} + 3 \cdot 2^{99} = 2^{99}.$$

□

Problem 4.165. Let a, b, c be positive numbers, such that $\log_2 a - 2 = \log_3 b = \log_5 c$ and $a + b = c$. What is the value of the product $a \cdot b \cdot c$?

(A) 1000   (B) 3600   (C) 100   (D) 625   (E) 4900

Solution. Answer. (B)
Let us assume that
$$\log_2 a - 2 = \log_3 b = \log_5 c = k.$$
Then, we have that
$$a = 2^{k+2}, b = 3^k, c = 5^k.$$

Given that
$$4 \cdot 2^k + 3^k = 5^k,$$
or equivalently
$$4 \cdot \left(\frac{2}{5}\right)^k + \left(\frac{3}{5}\right)^k = 1.$$
Note that the function
$$f(x) = 4 \cdot \left(\frac{2}{5}\right)^x + \left(\frac{3}{5}\right)^x$$
is decreasing, therefore given equation has no more than one solution. Moreover, note that $k = 2$ is a solution.
Thus, it follows that $abc = 2^4 \cdot 3^2 \cdot 5^2 = 3600$. $\square$

Problem 4.166. *How many eight-digit numbers are there which are divisible by 11 and are composed (without repetitions) only of digits 1, 2,..., 8?*

(A) 3456   (B) 1152   (C) 5040   (D) 40320   (E) 4608

Solution. Answer. (E)
Let $\overline{a_1 a_2 ... a_8}$ be 8-digit number with the given property. Then, according to the divisibility rule by 11, we have that
$$11 \mid (a_1 - a_2 + a_3 - a_4 + a_5 - a_6 + a_7 - a_8).$$
Note that
$$a_1 + a_2 + a_3 + a_4 + a_5 + a_6 + a_7 + a_8 = 36.$$
Thus, it follows that
$$11 \mid 2(18 - a_2 - a_4 - a_6 - a_8).$$
On the other hand
$$5 + 6 + 7 + 8 \geq a_2 + a_4 + a_6 + a_8 \geq 1 + 2 + 3 + 4.$$
Hence, we obtain that
$$a_2 + a_4 + a_6 + a_8 = 18.$$
Now, let us call numbers a and $9 - a$ "neighbors" and perform the following casework.
Case 1. There are neighbors among a_2, a_4, a_6, a_8.
Note that if we remove this neighbors, the remaining two numbers are found to be neighbors too. Therefore, the total number of such quadruples (a_2, a_4, a_6, a_8) is $\binom{4}{2} = 6$.
In this case, the total number of such eight-digit numbers is $6 \cdot 4! \cdot 4! = 3456$.
Case 2. There are no neighbors among a_2, a_4, a_6, a_8.
Then, we have that either
$$\{a_2, a_4, a_6, a_8\} = \{2, 3, 5, 8\},$$
or
$$\{a_2, a_4, a_6, a_8\} = \{1, 4, 6, 7\}.$$
In this case, the total number of such 8-digit numbers is $2 \cdot 4! \cdot 4! = 1152$.
Hence, in total there are $3456 + 1152 = 4608$ eight-digit numbers with such property. $\square$

Problem 4.167. *Let 2×10 rectangular grid be randomly covered by ten 1×2 rectangles (dominos). What is the probability that two cells in the fifth column from the left are covered by different dominos?*

(A) $\dfrac{64}{89}$ (B) $\dfrac{49}{89}$ (C) $\dfrac{25}{89}$ (D) $\dfrac{40}{89}$ (E) $\dfrac{1}{2}$

Solution. Answer. (B)
Let us denote by a_n the number of ways to cover $2 \times n$ rectangular grid by n dominos. We have that $a_1 = 1, a_2 = 2$ and $a_{n+2} = a_{n+1} + a_n, n = 1, 2, \ldots$. This is because there are only two ways to cover the left bottom square of $2 \times n$ rectangular grid.

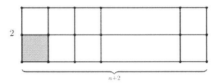

If the fifth column from the left is covered by different dominos, then these two cases are possible.

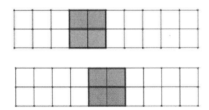

The probability that two cells in the fifth column from the left are covered by different dominos is:
$$p = \frac{a_3 \cdot a_5 + a_4 \cdot a_4}{a_{10}} = \frac{3 \cdot 8 + 5 \cdot 5}{89} = \frac{49}{89}.$$

□

Problem 4.168. *Let the vertices of triangle ABC lie on the surface of a sphere with a radius of 13 (see the figure). Given that $AB = 5$ and $m\angle ACB = 30°$. What is the value of the distance from the center of the sphere to the plane containing triangle ABC?*

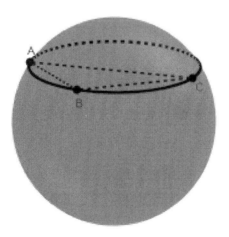

(A) 10 (B) 11 (C) 12 (D) 6 (E) 8

Solution. Answer. (C)
Let point O be the center of the sphere and point O_1 be the circumcenter of triange ABC (see the figure). Note that O_1 is not necessarily located inside of triangle ABC as it is shown in the figure.

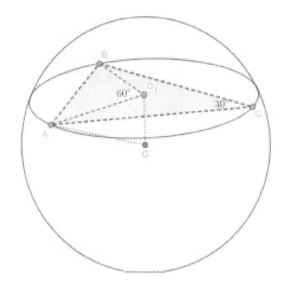

We have that
$$\angle AO_1B = 2\angle ACB = 60°,$$
therefore $\triangle ABO_1$ is an equilateral triangle. Thus, it follows that
$$AO_1 = AB = 5.$$
We have also $OO_1 \perp (ABC)$. Hence, we deduce that $\angle AO_1O = 90°$.
Therefore, from triangle AO_1O according to the Pythagorean theorem, we obtain that
$$OO_1 = \sqrt{AO^2 - AO_1^2} = 12.$$

□

Problem 4.169. *Let $f(x)$ be a monic polynomial function of degree five, such that*
$$f(2021) = -2, f(2022) = 1, f(2023) = 2, f(2024) = 1, f(2025) = -2.$$
What is the value of $f(2020)$?

(A) -5   (B) -12   (C) -127   (D) -210   (E) -2100

Solution. Answer. (C)
Let us consider the following function
$$h(x) = f(x) + (x - 2023)^2 - 2.$$
Note that $h(x)$ is a monic polynomial function of degree five too and
$$h(2021) = h(2022) = h(2023) = h(2024) = h(2025) = 0.$$
Thus, it follows that 2021, 2022, 2023, 2024, 2025 are roots of $h(x)$. We have that $h(x)$ is a monic polynomial function of degree five, therefore besides these five roots it does not have any other root. Hence, according to the linear factorization theorem we obtain that
$$h(x) = (x - 2021)(x - 2022)(x - 2023)(x - 2024)(x - 2025).$$

We deduce that
$$f(x) = (x-2021)(x-2022)(x-2023)(x-2024)(x-2025) - (x-2023)^2 + 2.$$

Thus, it follows that
$$f(2020) = (-1)(-2)(-3)(-4)(-5) - (-3)^2 + 2 = -120 - 9 + 2 = -127.$$

Alternative solution. According to Newton's divided difference formula, we have that
$$f(x) = (x-2021)(x-2022)(x-2023)(x-2024)(x-2025) + a(x-2021)(x-2022)(x-2023)(x-2024) +$$
$$+ b(x-2021)(x-2022)(x-2023) + c(x-2021)(x-2022) + d(x-2021) + e,$$

where a, b, c, d, e are some numbers.
Note that
$$\begin{cases} e = f(2021) = -2, \\ e + d = f(2022) = 1, \\ e + 2d + 2c = f(2023) = 2, \\ e + 3d + 6c + 6b = f(2024) = 1, \\ e + 4d + 12c + 24b + 24a = f(2025) = -2. \end{cases}$$

Solving this system of equations, we deduce that
$$e = -2, d = 3, c = -1, b = 0, a = 0.$$

Thus, it follows that
$$f(x) = (x-2021)(x-2022)(x-2023)(x-2024)(x-2025) - (x-2021)(x-2022) + 3(x-2021) - 2.$$

We obtain that
$$f(2020) = -120 - 2 - 3 - 2 = -127.$$

$\square$

Problem 4.170. *Let n be a positive integer and (a_n) be a number sequence defined as follows: $a_1 = a_2 = 1, a_3 = \dfrac{1}{2}$ and*
$$a_{n+1} = \frac{a_n^3 a_{n-2}}{a_{n-1}^3 + a_{n-2} a_{n-1} a_n},$$

where $n = 3, 4, \ldots$ What is the value of the following expression?
$$\frac{a_1}{a_2} + 2\frac{a_2}{a_3} + 3\frac{a_3}{a_4} + \ldots + 10\frac{a_{10}}{a_{11}}.$$

(A) 11!   (B) 11!-1   (C) 10!+3   (D) 10!+11   (E) $10 \cdot 11$

Solution. Answer. (B)
The given condition can be rewritten in the following way:
$$\frac{a_n^2}{a_{n-1} a_{n+1}} = \frac{a_{n-1}^2}{a_{n-2} a_n} + 1,$$

where $n = 3, 4, \ldots$ Let us consider the following sequence
$$b_n = \frac{a_{n+1}^2}{a_n a_{n+2}},$$

where $n = 1, 2, 3, \ldots$. We have that
$$b_1 = \frac{a_2^2}{a_1 a_3} = 2,$$
$$b_{n-1} = b_{n-2} + 1,$$
where $n = 3, 4, \ldots$. Thus, it follows that $b_n = n + 1$, where $n = 1, 2, \ldots$. We have that
$$\frac{a_2^2}{a_1 a_3} = 2, \frac{a_3^2}{a_2 a_4} = 3, \ldots, \frac{a_{n+1}^2}{a_n a_{n+2}} = n+1.$$

Multiplying all equations we obtain that
$$\frac{a_{n+1}}{a_{n+2}} = (n+1)!.$$

On the other hand, we have that
$$\frac{a_2}{a_3} = 2\frac{a_1}{a_2}, \frac{a_3}{a_4} = 3\frac{a_2}{a_3}, \ldots, \frac{a_{n+1}}{a_{n+2}} = (n+1)\frac{a_n}{a_{n+1}}.$$

Summing up the last equations we deduce that
$$S_n = \frac{a_1}{a_2} + 2\frac{a_2}{a_3} + 3\frac{a_3}{a_4} + \ldots + n\frac{a_n}{a_{n+1}} = (n+1)! - 1.$$

Thus, it follows that
$$S_{10} = 11! - 1.$$

$\square$

Problem 4.171. Let $z = \cos\frac{2\pi}{31} + i\sin\frac{2\pi}{31}$. What is the value of the following expression?
$$(1 - z + z^2)(1 - z^2 + z^4)(1 - z^4 + z^8)(1 - z^8 + z^{16})(1 - z^{16} + z^{32}).$$

(A) 1   (B) 0   (C) $\frac{1}{2}$ (D) $\frac{\sqrt{2}}{2}$ (E) 2^{31}

Solution. Answer. (A)
Note that
$$(1 - a + a^2)(1 + a + a^2) = (1 + a^2)^2 - a^2 = 1 + a^2 + a^4.$$

Thus, it follows that
$$1 - a + a^2 = \frac{1 + a^2 + a^4}{1 + a + a^2}.$$

Hence, we deduce that
$$(1 - z + z^2)(1 - z^2 + z^4)(1 - z^4 + z^8)(1 - z^8 + z^{16})(1 - z^{16} + z^{32}) =$$
$$= \frac{1 + z^2 + z^4}{1 + z + z^2} \cdot \frac{1 + z^4 + z^8}{1 + z^2 + z^4} \cdot \frac{1 + z^8 + z^{16}}{1 + z^4 + z^8} \cdot \frac{1 + z^{16} + z^{32}}{1 + z^8 + z^{16}} \cdot \frac{1 + z^{32} + z^{64}}{1 + z^{16} + z^{32}} = \frac{1 + z^{32} + z^{64}}{1 + z + z^2}.$$

Using de Moivre's formula, we obtain that
$$z^{31} = \left(\cos\frac{2\pi}{31} + i\sin\frac{2\pi}{31}\right)^{31} = \cos 2\pi + i\sin 2\pi = 1.$$

Thus, we deduce that
$$\frac{1 + z^{32} + z^{64}}{1 + z + z^2} = \frac{1 + z + z^2}{1 + z + z^2} = 1.$$

$\square$

Problem 4.172. *A league of soccer teams participate in a tournament. Any two teams play with each other exactly once. By the end of the tournament, exactly n games end in a draw and the total points gained in the tournament are equal to 2019. What is the value of the sum of all possible values of n? (Note: A winning team gets 3 points, a losing team gets 0 points, while a draw is 1 point for each team.)*

(A) 2019 (B) 2050 (C) 2150 (D) 4038 (E) 4080

Solution. Answer. (E)

Let m be the number of teams participating in the tournament. Thus, it follows that during the tournament in total $\frac{m(m-1)}{2}$ games were played. Given that

$$2n + 3\left(\frac{m(m-1)}{2} - n\right) = 2019,$$

or equivalently

$$n = \frac{3m(m-1)}{2} - 2019.$$

Note that $0 \leq n \leq \frac{m(m-1)}{2}$. Therefore $1346 \leq m(m-1) \leq 2019$. From the last double-inequality we obtain that $m \in \{38, 39, 40, 41, 42, 43, 44, 45\}$ and

$$n \in \left\{\frac{3(38^2 - 38)}{2} - 2019, \frac{3(39^2 - 39)}{2} - 2019, ..., \frac{3(45^2 - 45)}{2} - 2019\right\}.$$

Now, we need to calculate the following:

$$\frac{3}{2}(38^2 + 39^2 + ... + 45^2) - \frac{3}{2}(38 + 39 + ... + 45) - 8 \cdot 2019 =$$

$$= \frac{3}{2}\left(\frac{45 \cdot 46 \cdot 91}{6} - \frac{37 \cdot 38 \cdot 75}{6} - \frac{38 + 45}{2} \cdot 8\right) - 8 \cdot 2019 = 4080.$$

□

Problem 4.173. *Let vertices A, B, C of triangle ABC lie on circles with center O and radii $\sqrt{52}$, 3 and 5, respectively. Given that point O lies in the interior of triangle ABC and $\angle ABC = 120°$, $\angle BAC = 30°$. What is the value of the length of side AB?*

(A) 4 (B) 5 (C) 6 (D) 7 (E) 8

Solution. Answer. (D)

Note that

$$\angle BCA = 180° - 120° - 30° = 30°.$$

Thus, it follows that $AB = BC$.

Let us rotate clockwise triangle BOC around point B by $-120°$ (see the figure).

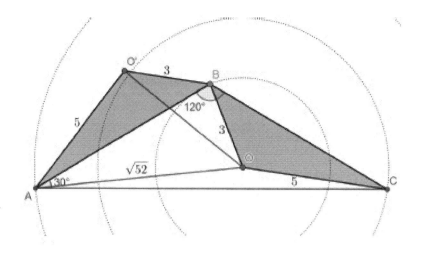

We have that
$$BO' = BO = 3,$$
and
$$\angle O'BO = 120°.$$
Hence, according to the law of cosines, from $\triangle O'BO$ we obtain that
$$OO' = \sqrt{3^2 + 3^2 - 2 \cdot 3 \cdot 3 \cdot \cos 120°} = 3\sqrt{3}.$$
From triangle $AO'O$ we have that
$$O'A^2 + O'O^2 = 25 + 27 = 52.$$
Hence, according to the converse theorem of the Pythagorean theorem, we obtain that
$$\angle AO'O = 90°.$$
On the other hand, we have that
$$\angle BO'O = \angle BOO' = \frac{180° - 120°}{2} = 30°.$$
Therefore, we deduce that
$$\angle AO'B = \angle AO'O + \angle BO'O = 90° + 30° = 120°.$$
From $\triangle AO'B$, according to the law of cosines, we obtain that
$$AB = \sqrt{3^2 + 5^2 - 2 \cdot 3 \cdot 5 \cdot \cos 120°} = 7.$$

□

Problem 4.174. Let n be a positive integer and (x_n) be a number sequence defined as follows: $x_1 = \dfrac{1}{\sqrt[6]{2}}$ and
$$\frac{x_n}{x_{n-1}^2} = 2^{1/2^n},$$
where $n = 2, 3, \ldots$. What is the value of the expression $(x_{2019})^{3 \cdot 2^{2019}}$?

(A) 2^{2019} (B) 2^{10} (C) 0.5 (D) 0.25 (E) 3

Solution. Answer. (C)
We have that
$$\frac{x_n}{x_{n-1}^2} = 2^{1/2^n}, \frac{x_{n-1}}{x_{n-2}^2} = 2^{1/2^{n-1}}, \ldots, \frac{x_2}{x_1^2} = 2^{1/2^2}.$$

Therefore, raising both sides of the second equation to 2^1, the third to 2^2, and so on, the last one to 2^{n-2} and multiplying all equations together, we obtain that
$$\frac{x_n}{x_1^{2^{n-1}}} = 2^{1/2^n + 1/2^{n-2} + \ldots + 2^{n-4}}.$$

Thus, it follows that
$$x_n = 2^{(-1/6) \cdot 2^{n-1}} \cdot 2^{(4^{n-1} - 1)/(3 \cdot 2^n)} = 2^{-1/(3 \cdot 2^n)}.$$

Hence, we deduce that
$$x_{2019} = 2^{-1/(3 \cdot 2^{2019})}.$$

We obtain that
$$(x_{2019})^{3 \cdot 2^{2019}} = 2^{-1} = 0.5.$$
$\square$

Problem 4.175. Let n be a positive integer, such that $\dfrac{(n!)^2}{(n+3)!}$ is also a positive integer. What is the smallest possible number of divisors of the expression $(n+1)(n+2)(n+3)$?

(A) 24 (B) 30 (C) 36 (D) 12 (E) 20

Solution. Answer. (B)
Given that
$$[(n+1)(n+2)(n+3)] \mid n!,$$
as we have that
$$\frac{(n!)^2}{(n+3)!} = \frac{n!}{(n+1)(n+2)(n+3)}.$$

Therefore, each of the numbers $n+1, n+2$, and $n+3$ is not prime. Let us do a casework:
Case 1. If n is even. Then any two of the numbers $n+1, n+2, n+3$ are relatively prime. Let
$$n+1 = ab, n+2 = cd, n+3 = ef.$$

Moreover
$$1 < a \leq b, 1 < c \leq d, 1 < e \leq f$$
and only one of the consecutive numbers $n+1, n+2, n+3$ can be a perfect square. Hence, if we replace each of the numbers a, b, c, d, e and f with their greatest prime divisor, we get that the number of divisors of the number $abcdef$ is not less than
$$2 \cdot 2 \cdot 2 \cdot 2 \cdot 3 = 48.$$

Case 2. If n is odd. Then one of the numbers $n+1$ and $n+3$ is even, while the other one is multiple of 4, larger than 4. Thus, it follows that

$$(n+1)(n+2)(n+3) = 8uvcd,$$

where each of the numbers u, v, c and d is greater than 1 and

$$(u,v) = 1, (u,c) = 1, (u,d) = 1, (v,c) = 1, (v,d) = 1,$$

c and d are odd numbers ($n+2 = cd$). Therefore, if we replace each of the numbers u, v, c and d with their greatest prime divisor we obtain that the number of divisors of $8uvcd$ is less than $5 \cdot 2 \cdot 3 = 30$. When $n = 7$, we have that

$$\frac{(n!)^2}{(n+3)!} = 7,$$

and

$$8 \cdot 9 \cdot 10 = 2^4 \cdot 3^2 \cdot 5$$

has $(4+1)(2+1)(1+1) = 30$ divisors. $\square$

4.8 Solutions of AMC 12 type practice test 8

Problem 4.176. *Some of the students in the class are students in an after-school math circle as well. Most of the students in the class participated in the AMC 12 test. It appeared that $\frac{7}{8}$ of the students atteding the math circle (from this class) and $\frac{5}{6}$ of the students not attending the math circle (from this class) participated in the AMC 12. Given that the total number of students in the class is not more than 19. What part of the class participated in the AMC 12?*

(A) $\frac{1}{2}$ (B) $\frac{2}{3}$ (C) $\frac{6}{7}$ (D) $\frac{3}{4}$ (E) $\frac{8}{9}$

Solution. Answer. (C)
Given that the number of the students attending the math circle in the class is a multiple of 8, while the number of students (in the class) not attending the math circle is a multiple of 6.
As the total number of the students in the class is not more than 19, then there are 8 students (in the class) attending the math circle and 6 students (in the class) not attending the math circle.
Thus, in the AMC 12 participated the following part of the class:
$$\frac{7+5}{8+6} = \frac{6}{7}.$$

$\square$

Problem 4.177. *Points $A(-1,4)$ and $B(2,10)$ lie on a rectangular coordinate plane. Point $C(x_0, y_0)$ lies on the segment AB, such that $AC : CB = 1 : 2$. What is the value of $x_0 + y_0$?*

(A) 6 (B) 5 (C) 4 (D) -6 (E) -5

Solution. Answer. (A)
Note that to reach point B from point A, one must shift 3 units to the right and 6 units up. Therefore, to point C from point A, one must shift 1 unit to the right and 2 units up, as $AC : AB = 1 : 3$. Thus, it follows that $x_0 = -1 + 1$ and $y_0 = 4 + 2$. Hence, we obtain that
$$x_0 + y_0 = 6.$$

$\square$

Problem 4.178. *Given that the mean of the numbers $30, 31, ..., 29 + 2n$ is equal to the mean of the numbers $100, 101, ..., 99 + n$. What is the value of the sum of all digits of n?*

(A) 10 (B) 5 (C) 8 (D) 7 (E) 6

Solution. Answer. (B)
We have that
$$30 + 29 + 2n = 100 + 99 + n.$$
Thus, it follows that $n = 140$. Therefore, the sum of all digits of n is $1 + 4 + 0 = 5$.

$\square$

Problem 4.179. *Given that the numbers $n!, (n+1)!, n!(n+19)$ form an arithmetic sequence, where n is a positive integer. What is the value of the product of the digits of n?*

(A) 1   (B) 24   (C) 30   (D) 3   (E) 8

Solution. Answer. (E)
We have that
$$n! + n!(n+19) = 2(n+1)!.$$
Thus, it follows that
$$1 + (n+19) = 2n + 2.$$
Hence, we obtain that $n = 18$. □

Problem 4.180. *Positive integers m and n are such that both the sum and the difference of $mn+m+n-53$ and $9n - 6m + 3$ are prime numbers. What is the value of the sum of all digits of the product mn?*

(A) 7   (B) 9   (C) 10   (D) 13   (E) 15

Solution. Answer. (B)
Given that the the sum and the difference of $mn + m + n - 53$ and $9n - 6m + 3$ are prime numbers. Therefore, we have that
$$mn - 5m + 10n - 50 = (m+10)(n-5),$$
and
$$mn + 7m - 8n - 56 = (m-8)(n+7),$$
are prime numbers. Since
$$n + 7 \geq 8 > 1,$$
and
$$m + 10 \geq 11 > 1,$$
then
$$m - 8 = 1,$$
and
$$n - 5 = 1.$$
Thus, it follows that $m = 9, n = 6$. Hence, we obtain that $mn = 54$.
Note that, in this case
$$\begin{cases} (m-8)(n+7) = 13, \\ (m+10)(n-5) = 19. \end{cases}$$
□

Problem 4.181. *Let the lengths of two of the sides of triangle ABC be 6, 8 and its area be 24. What is the value of the perimeter of triangle ABC?*

(A) 20   (B) 24   (C) 22   (D) 27   (E) 26

Solution. Answer. (B)
Let $AB = 6$ and $BC = 8$ (see the figure). We have that
$$Area(ABC) = \frac{1}{2} AB \cdot h_c,$$
or
$$h_c = 8 = BC.$$

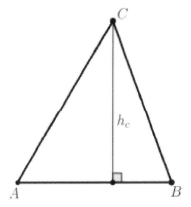

Thus, it follows that $\angle ABC = 90°$. According to the Pythagorean theorem, we obtain that
$$AC^2 = 6^2 + 8^2 = 10^2.$$
Hence, we deduce that $AC = 10$.
Therefore, the perimeter of $\triangle ABC$ is $6 + 8 + 10 = 24$. □

Problem 4.182. *Given*
$$f(x) = 2x^3 - 3x^2 + x + 5.$$
Let
$$A = f\left(\frac{1}{2020}\right) + f\left(\frac{2}{2020}\right) + f\left(\frac{3}{2020}\right) + \ldots + f\left(\frac{2018}{2020}\right) + f\left(\frac{2019}{2020}\right).$$
What is the value of the sum of all digits of A?

(A) 1 (B) 10 (C) 15 (D) 18 (E) 22

Solution. Answer. (C)
Let us evaluate the following sum $f(x) + f(1-x)$. We have that
$$f(x) + f(1-x) = 2x^3 - 3x^2 + x + 5 + 2(1-x)^3 - 3(1-x)^2 + 1 - x + 5 =$$
$$= 2(1 - 3x + 3x^2) - 3(2x^2 - 2x + 1) + 11 = 10.$$
Thus, it follows that
$$f(x) + f(1-x) = 10.$$
On the other hand, we have that
$$f\left(\frac{1}{2}\right) = 5.$$
Hence, we obtain that
$$f\left(\frac{1}{2020}\right) + f\left(\frac{2}{2020}\right) + f\left(\frac{3}{2020}\right) + \ldots + f\left(\frac{2018}{2020}\right) + f\left(\frac{2019}{2020}\right) =$$
$$= \left(f\left(\frac{1}{2020}\right) + f\left(\frac{2019}{2020}\right)\right) + \left(f\left(\frac{2}{2020}\right) + f\left(\frac{2018}{2020}\right)\right) + \ldots + \left(f\left(\frac{1009}{2020}\right) + f\left(\frac{1011}{2020}\right)\right) + f\left(\frac{1010}{2020}\right) =$$
$$= 10 + 10 + \ldots + 10 + 10 + 5 = 1009 \cdot 10 + 5 = 10095.$$
Therefore, the sum of all digits is $1 + 0 + 0 + 9 + 5 = 15$. □

Problem 4.183. *For how many positive integers n the values of fractions $\dfrac{20n+19}{n+817}$ and $\dfrac{19n+61}{n+817}$ are positive integers too?*

(A) 0   (B) 1   (C) 3   (D) 5   (E) 11

Solution. Answer. (B)
Given that
$$\frac{20n+19}{n+817} - \frac{19n+61}{n+817} = \frac{n-42}{n+817}.$$
is an integer. Thus, it follows that $n = 42$. Hence, we deduce that
$$\frac{20n+19}{n+817} = 1,$$
and
$$\frac{19n+61}{n+817} = 1.$$
Therefore, there is only one such value of n. □

Problem 4.184. *What is the smallest possible integer value of x, such that $\log_2 x, \log_4 x, \log_2 \dfrac{x}{4}$ are the lengths of sides of the same triangle?*

(A) 5   (B) 10   (C) 16   (D) 17   (E) 18

Solution. Answer. (D)
Given that $\log_2 x, \log_4 x$ and $\log_2 \dfrac{x}{4}$ are the lengths of sides of the same triangle. Thus, it follows that
$$\begin{cases} \log_2 \dfrac{x}{4} > 0, \\ \log_2 x < \log_4 x + \log_2 \dfrac{x}{4} = \dfrac{1}{2}\log_2 x + \log_2 x - 2. \end{cases}$$
Hence, we deduce that $x > 16$. Thus, the smallest possible integer value of x satisfying the assumptions of the problem is equal to 17. □

Problem 4.185. *A hotel with infinite number of rooms has room numbers labeled 1, 2, 3,... Given that a guest is assigned to room i with probability $\dfrac{1}{2^i}$. What is the probability that the positive difference of the room numbers assigned to two random guests is 2?*

(A) $\dfrac{1}{3}$ (B) $\dfrac{2}{3}$ (C) $\dfrac{1}{12}$ (D) $\dfrac{1}{6}$ (E) $\dfrac{1}{4}$

Solution. Answer. (D)
Note that the probability of the event that the first guest is assigned the room i (where $i \geq 3$) and the second guest is assigned either the room $i-2$ or $i+2$ equals to
$$\frac{1}{2^i} \cdot \frac{1}{2^{i-2}} + \frac{1}{2^i} \cdot \frac{1}{2^{i+2}}.$$
Therefore, the required probability is
$$\frac{1}{2} \cdot \frac{1}{2^3} + \frac{1}{2^2} \cdot \frac{1}{2^4} + \left(\frac{1}{2^3} \cdot \frac{1}{2} + \frac{1}{2^3} \cdot \frac{1}{2^5}\right) + \left(\frac{1}{2^4} \cdot \frac{1}{2^2} + \frac{1}{2^4} \cdot \frac{1}{2^6}\right) + \ldots =$$
$$= \frac{2}{2 \cdot 2^3} + \frac{2}{2^2 \cdot 2^4} + \frac{2}{2^3 \cdot 2^5} + \ldots = \frac{1}{2^3} + \frac{1}{2^5} + \ldots = \frac{\dfrac{1}{8}}{1 - \dfrac{1}{4}} = \frac{1}{6}.$$
□

Problem 4.186. *What is the total number of all unordered pairs of edges of a regular octahedron, such that no plane can pass through them?*

(A) 12   (B) 24   (C) 36   (D) 42   (E) 48

Solution. Answer. (B)
Note that for each edge of a regular octahedron there are exactly 4 pairs of skew edges (recall that in three-dimensional space skew lines are two lines that do not intersect and are not parallel).

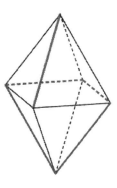

Therefore, in total there are $\dfrac{12 \cdot 4}{2} = 24$ unordered pairs of skew edges.   □

Problem 4.187. *A circle σ of radius $2 - \sqrt{3}$ and line l tangent to circle σ at point B lie in the same plane. Let A be a point outside circle σ and AM be tangent to circle σ, such that it does not intersect with line l. Given that $AM = 1$. Let C, D be points on line l, such that $BC = BD = 1$. What is the angle measure (in degrees) of $\angle CAD$?*

(A) 15   (B) 30   (C) 45   (D) 60   (E) 75

Solution. Answer. (E)
Let O be the center of circle σ (see the figure).
Note that the corresponding legs of right triangles AMO, CBO, DBO are equal to each other, therefore these triangles are pairwise congruent to each other. Thus, it follows that

$$AO = CO = DO.$$

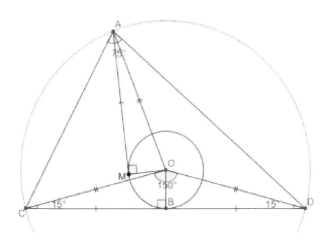

If $\angle OCB = \alpha$, then from right triangle CBO, we have that
$$\tan \alpha = \frac{OB}{CB} = 2 - \sqrt{3}.$$
Taking this into consideration, we obtain that
$$\tan \alpha + \tan 60° = 2.$$
Thus, it follows that
$$\frac{\sin(\alpha + 60°)}{\cos \alpha \cdot \cos 60°} = 2.$$
Hence, we deduce that
$$\sin(\alpha + 60°) = \sin(90° - \alpha).$$
Therefore
$$\alpha + 60° = 90° - \alpha.$$
We obtain that $\alpha = 15°$.
Thus, it follows that $\angle COD = 150°$ and $\angle CAD = 75°$. $\square$

Problem 4.188. *Let a be the smallest possible value of x, such that the values of the functions $y = \frac{5}{6}x + \frac{1}{3}$ and $y = \frac{25}{16}x - \frac{1}{2}$ are positive integers. What is the value of the sum of all digits of a^3?*

(A) 1   (B) 2   (C) 3   (D) 6   (E) 8

Solution. Answer. (E)
Let
$$\frac{5}{6}a + \frac{1}{3} = m,$$
and
$$\frac{25}{16}a - \frac{1}{2} = k,$$
where m and k are positive integers. We need to find the least possible value of m. We have that
$$a = \frac{6m - 2}{5},$$
and
$$a = \frac{16}{25}k + \frac{8}{25}.$$
Hence, we deduce that
$$\frac{6m - 2}{5} = \frac{16}{25}k + \frac{8}{25},$$
or equivalently
$$30m - 10 = 16k + 8.$$
Therefore
$$15m - 9 = 8k.$$
Thus, it follows that the solutions are
$$m = 7 + 8q, k = 12 + 15q,$$
where q is an nonnegative integer.
Hence, the least possible value of m is 7. We obtain that $a = 8$ and $a^3 = 512$. Thus, the sum of all digits of a^3 is equal to 8. $\square$

Problem 4.189. Let point $M(x_0, y_0)$ and circle σ lie in the same plane. Let MB and MD be secants, such that rays BM and DM intersect σ at points A and C, respectively. Given that $AB = CD$ and $A(2, 4), B(3, 6), C(0, 2)$. What is the value of $4x_0 + y_0$?

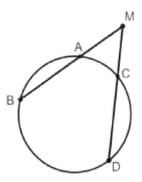

(A) 2   (B) 3   (C) 8   (D) 0   (E) 5

Solution. Answer. (C)
Let O be the center of circle σ. From the condition $AB = CD$ it follows that $\triangle OAB$ is congruent to $\triangle OCD$. Hence, we obtain that O is equidistant from the sides of $\angle BMD$. Therefore $MA = MC$. Note that points A and B lie on the line $y = 2x$, so M also lies on it.
Thus, it follows that
$$\begin{cases} y_0 = 2x_0, \\ \sqrt{(x_0 - 2)^2 + (y_0 - 4)^2} = \sqrt{(x_0 - 0)^2 + (y_0 - 2)^2}. \end{cases}$$

Hence, we deduce that
$$\begin{cases} y_0 = 2x_0, \\ x_0 + y_0 = 4. \end{cases}$$

Therefore, we get that
$$x_0 = \frac{4}{3}, y_0 = \frac{8}{3}.$$

Thus, it follows that
$$4x_0 + y_0 = 4 \cdot \frac{4}{3} + \frac{8}{3} = 8.$$

$\square$

Problem 4.190. *What is the probability that a random positive divisor of 2020^5 is not ending in 0?*

(A) $\dfrac{8}{33}$ (B) $\dfrac{2}{11}$ (C) $\dfrac{2}{3}$ (D) $\dfrac{3}{4}$ (E) $\dfrac{10}{101}$

Solution. Answer. (A)
Let us write prime factorization of 2020^5. We have that
$$2020^5 = 2^{10} \cdot 5^5 \cdot 101^5.$$

Therefore, the number of all possible divisors of 2020^5 is $11 \cdot 6 \cdot 6 = 396$.
Now, let us find the number of all divisors of 2020^5 ending in 0. All such divisors are of the following form
$$2^\alpha \cdot 5^\beta \cdot 101^\gamma,$$

where
$$\alpha \in \{1, 2, ..., 10\}, \beta \in \{1, 2, ..., 5\}, \gamma \in \{0, 1, ..., 5\}.$$

Therefore, in total there are $10 \cdot 5 \cdot 6 = 300$ such divisors.

Thus, it follows that the required probability is
$$\frac{396 - 300}{396} = \frac{96}{396} = \frac{8}{33}.$$

□

Problem 4.191. *What is the total number of all complex numbers z, such that $0, z, z^2$ are the vertices of an isosceles right triangle?*

(A) 0   (B) 2   (C) 4   (D) 6   (E) 8

Solution. Answer. (D)

Let us perform a casework with respect to $|z|$.

Case 1. If $|z| = 1$.

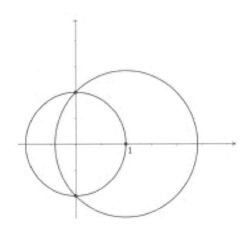

In this case, we have that
$$|z^2| = |z|^2 = 1.$$

Thus, it follows that
$$|z^2 - z| = \sqrt{2}.$$

Taking this into consideration and as $|z| = 1$, we obtain that
$$|z - 1| = \sqrt{2}.$$

Therefore, in this case there are two possible answers.

Case 2. If $|z| > 1$.

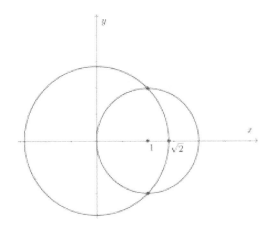

In this case, we have that
$$|z^2| > |z|.$$
Thus, it follows that
$$|z| = |z^2 - z|.$$
We deduce that
$$|z^2| = \sqrt{2}|z|.$$
Hence, we obtain that
$$|z| = \sqrt{2},$$
and
$$|z - 1| = 1.$$
Therefore, in this case also there are two possible answers.

Case 3. If $|z| < 1$.

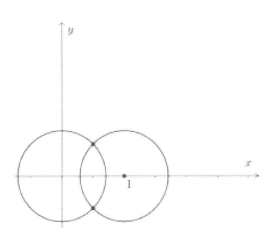

Then, we have that
$$|z^2| < |z|.$$

Thus, it follows that
$$|z^2| = |z^2 - z|.$$
We deduce that
$$\sqrt{2}|z^2| = |z|.$$
Hence, we obtain that
$$|z| = \frac{\sqrt{2}}{2},$$
and
$$|z - 1| = \frac{\sqrt{2}}{2}.$$
Therefore, in this case also there are two possible answers.

Thus, in total there are $2 + 2 + 2 = 6$ such complex numbers. □

Problem 4.192. *Let ABC be a triangle, such that $AC = 3, BC = 5, AB = 7$ and its each side is the diameter of a semicircle (see the figure). Let S_1, S_2, S_3 be the areas of the grey parts. What is the value of $S_1 + S_3 - S_2$?*

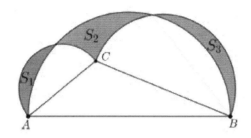

(A) $\dfrac{15}{8}(2\sqrt{3} - \pi)$ (B) $\dfrac{15}{4}(2\sqrt{3} - \pi)$ (C) $\dfrac{\pi}{16} - \dfrac{\sqrt{3}}{18}$ (D) 4 (E) $\pi + 2$

Solution. Answer. (A)

Note that
$$S_1 + S_3 + \frac{\pi \cdot 3.5^2}{2} = \frac{\pi \cdot 1.5^2}{2} + \frac{\pi \cdot 2.5^2}{2} + S_2 + Area(ABC).$$

According to Heron's formula, we have that
$$Area(ABC) = \sqrt{\frac{15}{2} \cdot \frac{1}{2} \cdot \frac{9}{2} \cdot \frac{5}{2}} = \frac{15\sqrt{3}}{4}.$$

Thus, it follows that
$$S_1 + S_3 - S_2 = -\frac{\pi}{2}(3.5^2 - 1.5^2 - 2.5^2) + \frac{15\sqrt{3}}{4} = \frac{15}{8}(2\sqrt{3} - \pi).$$

□

Problem 4.193. *Let A be the smallest positive ten-digit number that is divisible by 36 and whose base 10 representation consists of only 4's and 9's, with at least one of each. What is the remainder of A after division by 11?*

(A) 1 (B) 3 (C) 5 (D) 8 (E) 10

Solution. Answer. (C)
This number A is divisible by 36, therefore $9 \mid A$ and $4 \mid A$.
From the condition $4 \mid A$ it follows that the last two digits have to be divisible by 4. That means the last two digits of this integer are 4 and 4.
From the condition $9 \mid A$ it follows that the sum of all digits is divisible by 9. Given that A is a ten-digit number and its last two digits are 4's, the total number of all other 4's in A (except these two 4's) we denote by n, then for the sum of all digits of A we obtain that:

$$9 \mid (4 + 4 + 4 \cdot n + 9 \cdot (8 - n)).$$

We deduce that $9 \mid (1 + 5n)$. Thus, it follows that $n = 7$. Hence, we have nine 4's and one 9.
Note that the smallest positive ten-digit number that satisifes all these conditions is 4444444944.
The remainder of 4444444944 after division by 11 is equal to 5, as

$$4444444944 = 4444444444 + 495 + 5 = 11 \cdot 404040404 + 11 \cdot 45 + 5.$$

Therefore, the remainder of A after division by 11 is equal to 5. $\square$

Problem 4.194. *Let the base of pyramid $SABCD$ be a square with side length 4. Given that $SA = 3$, where SA is the altitude to the base. M is a point inside the pyramid equidistant from all its faces. Denote by h the distance from point M to any of its faces. What is the value of h?*

(A) 1 (B) 2 (C) 3 (D) 4 (E) 5

Solution. Answer. (A)
We have that $BC \perp AB$ and $SA \perp BC$. Thus, it follows that BC is perpendicular to plane SAB (see the figure). Let N be the projection of M on plane SAB. As $MN \parallel BC$ and line MN does not belong to planes SAD, SBC, ABC, then the distance from point M to faces SAD, SBC, ABC is equal to the distance from point N to planes SAD, SBC, ABC. We have that N is equidistant from lines SA, AB, SB. Therefore, N is the incenter of triangle SAB.

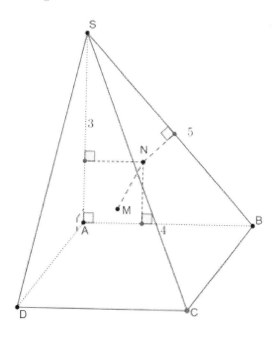

On the other hand, using the formula of the inradius of a right triangle, we obtain that the inradius r of right triangle SAB is:
$$r = \frac{SA + AB - SB}{2}.$$
Hence, the distance h from point M to any of the faces of the pyramid is:
$$h = \frac{3 + 4 - 5}{2} = 1.$$

□

Problem 4.195. *For how many quadratic trinomials $x^2 + px + q$ each of the numbers $p + q + 1$ and $p - q + 1$ is a root of given quadratic trinomial?*

(A) 0 (B) 1 (C) 4 (D) 2 (E) 3

Solution. Answer. (C)
Let us perform a casework with respect to $p + q + 1$.
Case 1. If $p + q + 1 \neq p - q + 1$.
Then, according to Vieta's formula, we have that
$$\begin{cases} p + q + 1 + p - q + 1 = -p, \\ (p + q + 1)(p - q + 1) = q. \end{cases}$$
Thus, it follows that
$$p = -\frac{2}{3}, q^2 + q - \frac{1}{9} = 0.$$
Therefore, in this case there are 2 possible pairs (p, q).
Case 2. If $p + q + 1 = p - q + 1$.
Then, $q = 0$ and given that
$$(p + 1)^2 + p(p + 1) = 0.$$
We deduce that
$$(p + 1)(2p + 1) = 0.$$
Hence, we obtain that either $p = -1$ or $p = -\frac{1}{2}$.
Therefore, in this case too there are 2 possible pairs (p, q).
Thus, in total there are 4 possible pairs (p, q).

□

Problem 4.196. *A random four-digit number is chosen. What is the probability that the digits of the chosen number are four consecutive numbers?*

(A) $\frac{1}{30}$ (B) $\frac{9}{500}$ (C) $\frac{7}{300}$ (D) $\frac{11}{450}$ (E) $\frac{7}{80}$

Solution. Answer. (B)
Let us find the number of four-digit numbers with consecutive digits.
The number of four-digit numbers with digits 0, 1, 2, 3 is $3 \cdot 3 \cdot 2 \cdot 1 = 18$.
The number of four-digit numbers with digits 1, 2, 3, 4 is $4 \cdot 3 \cdot 2 \cdot 1 = 24$.
And so on, the number of four-digit numbers with digits 6, 7, 8, 9 is $4 \cdot 3 \cdot 2 \cdot 1 = 24$.
Therefore, the total number of four-digit numbers with the above mentioned property is $6 \cdot 24 + 18 = 162$.
In total, there are $9 \cdot 10 \cdot 10 \cdot 10 = 9000$ four digit numbers.
Hence, the required probability is the following:
$$\frac{162}{9000} = \frac{9}{500}.$$

□

Problem 4.197. *Let ABCD be a convex quadrilateral, such that $AD = 1, CD = 2, \angle ABC = 90°$ and $AB : BC = 3 : 4$. What is the greatest possible value of the area of ABCD?*

(A) $\dfrac{30 + \sqrt{1224}}{25}$ (B) 2 (C) $\dfrac{30 + \sqrt{1201}}{25}$ (D) 2.5 (E) 3

Solution. Answer. (C)
Let $AB = 3x, \angle DAC = \alpha$ and $\angle DCA = \beta$. Given that $BC = 4x$ and $AC = 5x$ (see the figure).

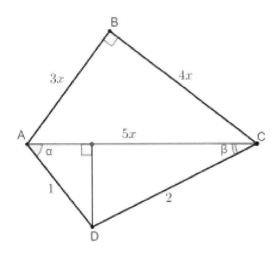

We have $\sin\alpha = 2\sin\beta$ and $5x = \cos\alpha + 2\cos\beta$.
Thus, it follows that

$$Area(ABCD) = Area(ABC) + Area(ADC) = 6x^2 + \sin(\alpha + \beta) = \dfrac{6}{25}(\cos\alpha + 2\cos\beta)^2 + \sin(\alpha + \beta) =$$

$$= \dfrac{6}{25}(\cos\alpha + 2\cos\beta)^2 + \dfrac{6}{25}(\sin\alpha - 2\sin\beta)^2 + \sin(\alpha + \beta) =$$

$$= \dfrac{6}{25}(\cos^2\alpha + \sin^2\alpha) + \dfrac{24}{25}(\cos^2\beta + \sin^2\beta) + \dfrac{24}{25}\cos(\alpha+\beta) + \sin(\alpha+\beta) =$$

$$= \dfrac{6}{5} + \dfrac{1}{25}(25\sin(\alpha+\beta) + 24\cos(\alpha+\beta)) = \dfrac{6}{5} + \dfrac{\sqrt{1201}}{25}\sin(\alpha+\beta+\phi) \leq \dfrac{6}{5} + \dfrac{\sqrt{1201}}{25}.$$

We obtain that
$$Area(ABCD) \leq \dfrac{30 + \sqrt{1201}}{25}.$$

Note that the equality holds true (see the figure) under the following conditions

$$\begin{cases} \sin(\alpha+\beta) = \dfrac{25}{\sqrt{1201}}, \\ \cos(\alpha+\beta) = \dfrac{24}{\sqrt{1201}}, \\ \sin\alpha = 2\sin\beta. \end{cases}$$

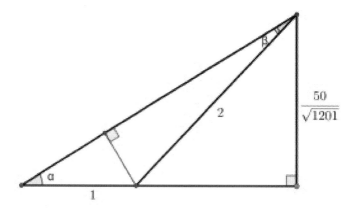

Therefore, the greatest possible value of the area of $ABCD$ is equal to $\dfrac{30+\sqrt{1201}}{25}$. □

Problem 4.198. *How many ten-digit numbers $\overline{a_1a_2...a_{10}}$ with nonzero digits exist, such that each of the three-digit numbers $\overline{a_1a_2a_3}, \overline{a_2a_3a_4}, ..., \overline{a_8a_9a_{10}}$ is not divisible by 3?*

(A) $2^8 \cdot 3^{10}$ (B) $2^8 \cdot 3^{12}$ (C) $2^{10} \cdot 3^{10}$ (D) $2^{12} \cdot 3^{12}$ (E) $2^{12} \cdot 3^8$

Solution. Answer. (B)
Note that both a_1 and a_2 can be chosen in 9 ways, while each of the numbers $a_3, ..., a_{10}$ in 6 ways. Therefore, the required total number of 10-digit numbers is

$$9 \cdot 9 \cdot 6 \cdot ... \cdot 6 = 3^4 \cdot 6^8 = 3^{12} \cdot 2^8.$$

□

Problem 4.199. *Let n be a positive integer and sequence (x_n) be defined as follows: $x_1 = 1$ and $x_{n+1} = x_n + \dfrac{1}{x_n}$, where $n = 1, 2, ...$ What is the value of the sum of all digits of the smallest possible value of n, such that $x_n > 8$?*

(A) 5 (B) 12 (C) 16 (D) 21 (E) 24

Solution. Answer. (A)
Taking the square on both sides of given equation, we obtain that

$$x_{i+1}^2 = x_i^2 + 2 + \dfrac{1}{x_i^2},$$

where $i = 1, 2, ..., n$. Summing up all equations (for $i = 1, 2, ..., n$) we obtain that

$$x_{n+1}^2 = 2n + 2 + \dfrac{1}{x_2^2} + ... + \dfrac{1}{x_n^2}.$$

Thus, it follows that

$$x_{n+1}^2 > 2n + 2,$$

where $n = 2, 3, ...$ We deduce that

$$x_{32} > \sqrt{2 \cdot 31 + 2} = 8.$$

On the other hand, we have that

$$x_{31}^2 = 62 + \frac{1}{x_2^2} + \frac{1}{x_3^2} + \ldots + \frac{1}{x_{30}^2} < 62 + \frac{1}{4} + \frac{1}{6} + \ldots + \frac{1}{60} <$$

$$< 62 + \frac{1}{2}\left(\left(\frac{1}{2} + \frac{1}{3}\right) + \left(\frac{1}{4} + \ldots + \frac{1}{7}\right) + \left(\frac{1}{8} + \ldots + \frac{1}{15}\right) + \left(\frac{1}{16} + \ldots + \frac{1}{31}\right)\right) <$$

$$< 62 + \frac{1}{2}\left(2 \cdot \frac{1}{2} + 4 \cdot \frac{1}{4} + 8 \cdot \frac{1}{8} + 16 \cdot \frac{1}{16}\right) = 64.$$

Hence, we deduce that $x_{31} < 8$.
Therefore, the smallest possible value of n such that $x_n > 8$ is equal to 32. Hence, the sum of all digits of the smallest possible value of n is equal to 5. □

Problem 4.200. *Points $O(0,0), A(1,1), B(2,3), C(-1,2), D(-2,5)$ are drawn on a rectangular coordinate plane. Let M, N be any points on line segments AB, CD, respectively. What is the area of a figure formed by the locus of all points X, such that $\overrightarrow{OX} = \overrightarrow{OM} + \overrightarrow{ON}$?*

   (A) $3+2\sqrt{2}$ (B) 6   (C) 4.5   (D) $3\sqrt{3}$ (E) 5

Solution. Answer. (E)
Let us consider point K (see the figure), such that

$$\overrightarrow{OK} = \frac{1}{2}(\overrightarrow{OM} + \overrightarrow{ON}).$$

It is easy to note that K is the midpoint of segment MN as well as segment OX.

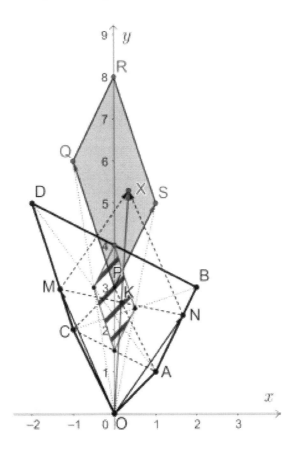

Hence, the requested locus of points X is the parallelogram $PQRS$, where
$$\overrightarrow{OP} = \overrightarrow{OA} + \overrightarrow{OC}, \overrightarrow{PQ} = \overrightarrow{CD}, \overrightarrow{PS} = \overrightarrow{AB}.$$

Let
$$\angle(\overrightarrow{AB}, \overrightarrow{CD}) = \alpha.$$

Thus, it follows that
$$Area(PQRS) = PQ \cdot RS \cdot \sin\alpha = AB \cdot CD \cdot \sqrt{1 - \cos^2\alpha} =$$
$$= AB \cdot CD \cdot \sqrt{1 - \left(\frac{\overrightarrow{AB} \cdot \overrightarrow{CD}}{AB \cdot CD}\right)^2} = \sqrt{(AB \cdot CD)^2 - (\overrightarrow{AB} \cdot \overrightarrow{CD})^2}.$$

We have that
$$\overrightarrow{AB} = (1, 2), AB^2 = 5,$$
and
$$\overrightarrow{CD} = (-1, 3), CD^2 = 10.$$

Hence, we obtain that
$$\overrightarrow{AB} \cdot \overrightarrow{CD} = (-1) \cdot 1 + 2 \cdot 3 = 5.$$

Thus, it follows that
$$(\overrightarrow{AB} \cdot \overrightarrow{CD})^2 = 25.$$

We deduce that
$$(AB \cdot CD)^2 = 5 \cdot 10 = 50.$$

Taking into consideration the above mentioned equations, we obtain that
$$Area(PQRS) = \sqrt{50 - 25} = 5.$$

□

4.9 Solutions of AMC 12 type practice test 9

Problem 4.201. *What is the value of the following expression?*
$$(2^{-1} + 3^{-1} - 4^{-1})^{-1}.$$

(A) $\dfrac{7}{12}$ (B) 12   (C) $\dfrac{12}{7}$ (D) 1   (E) 9

Solution. Answer. (C)
$$(2^{-1} + 3^{-1} - 4^{-1})^{-1} = \left(\frac{1}{2} + \frac{1}{3} - \frac{1}{4}\right)^{-1} = \left(\frac{7}{12}\right)^{-1} = \frac{12}{7}.$$

□

Problem 4.202. *David and Ann have together some candies. Assume that David gives half of his candies to Ann, afterward Ann gives half of her candies to David, then David gives half of his candies to Ann, and finally Ann gives half of her candies to David. In the end, David and Ann have 15 and 9 candies, respectively. How many candies did Ann have in the beginning?*

(A) 24   (B) 20   (C) 16   (D) 8   (E) 10

Solution. Answer. (A)
Let us start from backward.

David	Ann
15	9
6	18
12	12
0	24
0	24

Therefore, the correct answer is 24.

□

Problem 4.203. *A factory that produces cars overperformed the plan for January by 20% and underperformed the plan for February by 25%. It turned out, that in total the factory exactly performed their two months plan. By how many percents is the plan for February less than the plan for January?*

(A) 25   (B) 45   (C) 20   (D) 10   (E) 5

Solution. Answer. (C)
Assume that according to the plan of January the factory needed to produce x cars and according to the plan of February the factory needed to produce y cars. From the conditions of the problem, it follows that
$$\frac{120}{100}x + \frac{75}{100}y = x + y.$$
Hence, we obtain that
$$y = \frac{4}{5}x = \frac{80}{100}x.$$
Therefore, the answer is 20%.

□

Problem 4.204. *Given that the sum of two consecutive odd prime numbers is not divisible by 4. What is the smallest possible number of divisors that this sum may have?*

(A) 8   (B) 9   (C) 4   (D) 5   (E) 6

Solution. Answer. (E)
Let p and q be considered odd prime consecutive numbers. As p and q are odd numbers, therefore $p + q = 2n$. Note that n is an odd number in between p and q, thus n is a composite number (as p and q are consecutive primes and n is in between, so n cannot be prime). Thus, the numbers of divisors of $2n$ is not less than 6. Note that $7 + 11 = 18$ and 18 has 6 divisors. Therefore, the answer is 6. □

Problem 4.205. *There were some candies on the table. Alexa ate half of the candies and a half of one candy. Afterward, her brother ate half of the remaining candies and a half of one candy. Finally, there is one candy left on the table. How many candies were there on the table in the beginning?*

(A) 15   (B) 9   (C) 13   (D) 11   (E) 7

Solution. Answer. (E)
Let the initial number of candies be x. Alexa ate $\frac{x}{2} + \frac{1}{2}$ candies and her brother ate $\frac{x-1}{4} + \frac{1}{2}$ candies. Given that
$$\frac{x}{2} + \frac{1}{2} + \frac{x-1}{4} + \frac{1}{2} + 1 = x.$$
Hence, it follows that $x = 7$. □

Problem 4.206. *What is the total number of negative solutions of the following equation?*
$$(1+x)(1+x^2)(1+x^4) = 1.$$

(A) 5   (B) 4   (C) 3   (D) 0   (E) 1

Solution. Answer. (D)
The given equation can be rewritten as
$$(1-x)(1+x)(1+x^2)(1+x^4) = 1 - x.$$
We deduce that
$$1 - x^8 = 1 - x.$$
Thus, it follows that
$$x(x^7 - 1) = 0.$$
We obtain that, either $x = 0$ or $x = 1$.
Therefore, given equation does not have any negative solutions. Hence, the answer is 0. □

Problem 4.207. *In triangle ABC points D, E are on sides AB, BC, respectively. Let circumcenter O of triangle ABC be the intersection point of segments CD and AE. Given that $BD = BO = BE$ and $\angle ABC = n°$. What is the value of n?*

(A) 72   (B) 60   (C) 30   (D) 36   (E) 45

Solution. Answer. (A)
Let $\angle OAB = \alpha, \angle OCB = \beta, \angle OAC = \gamma$ (see the figure). From isosceles triangles AOB and OBD we have that $\angle OBA = \angle OAB = \alpha$ and $\angle BDO = 90° - \frac{\alpha}{2}$.

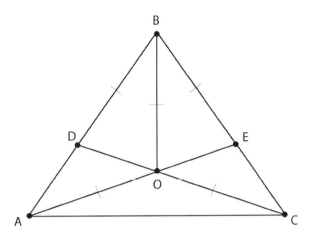

In a similar way, we obtain that
$$\angle OBC = \angle OCB = \beta,$$
and
$$\angle BEO = 90° - \frac{\beta}{2}.$$

From $\triangle ACD$ and $\triangle ACE$ according to the exterior angle theorem, it follows that
$$90° - \frac{\alpha}{2} = \alpha + \gamma + \gamma,$$
and
$$90° - \frac{\beta}{2} = \beta + \gamma + \gamma.$$

Therefore $\alpha = \beta$. We deduce that
$$3\alpha + 4\gamma = 180°.$$

On the other hand, we have that
$$4\alpha + 2\gamma = 180°.$$

Hence, we obtain that $\alpha = 2\gamma, \gamma = 18°$.
Thus, it follows that
$$\angle ABC = 2\alpha = 72°.$$

Therefore $n = 72$. □

Problem 4.208. *John has cut with a knife (by straight lines) each of the identical chocolate bars into two pieces, afterward he has cut (by straight lines) some of the pieces into two pieces and finished his actions. Given that every time John performed the cutting action he selected either exactly one chocolate bar to cut or exactly one piece to cut. Given also that John performed the cutting action 19 times and in the end he had 28 pieces. What is the total number of chocolate bars?*

(A) 7 (B) 4 (C) 14 (D) 9 (E) 8

Solution. Answer. (D)
Let the number of chocolate bars be n.
Note that every time John performs the cutting action the total number of pieces increases by 1.
Thus, if John performs the cutting action 19 times than the total number of pieces is equal to $n + 19$.
According to the condition of the problem, we have that
$$n + 19 = 28.$$

Hence $n = 9$. Therefore, the total number of chocolate bars is equal to 9. □

Problem 4.209. Let x, y be real numbers, such that $\sqrt{x} - \sqrt{x - 2019} = \sqrt{y + 2019} - \sqrt{y}$. What is the value of $x - y$?

(A) 1   (B) 100   (C) 2019   (D) -2019   (E) 1001

Solution. Answer. (C)
We have that
$$\sqrt{x} + \sqrt{y} = \sqrt{x - 2019} + \sqrt{y + 2019}.$$
Thus, it follows that
$$(\sqrt{x} + \sqrt{y})^2 = (\sqrt{x - 2019} + \sqrt{y + 2019})^2.$$
Hence, we obtain that
$$xy = (x - 2019)(y + 2019).$$
We deduce that
$$2019x - 2019y - 2019^2 = 0.$$
Therefore $x - y = 2019$. □

Problem 4.210. Let each of the areas of congruent equilateral triangles ABC and DEF be equal to 16 (see the figure). Given that
$$AE + AK = EB + BC + CK.$$
What is the value of the area of triangle AEK?

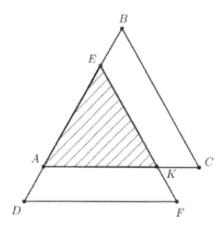

(A) 5   (B) 4   (C) 8   (D) 9   (E) 10

Solution. Answer. (D)
Note that if $AK = \lambda \cdot AC$ (see the figure), then
$$\lambda \cdot AC + \lambda \cdot AC = (1 - \lambda) \cdot AC + (1 - \lambda) \cdot AC + AC.$$
Thus, it follows that $\lambda = \dfrac{3}{4}$.

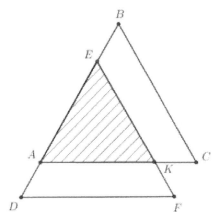

We have that
$$\frac{Area(AEK)}{Area(ABC)} = \frac{AK^2}{AC^2}.$$

Therefore $Area(AEK) = 9$. □

Problem 4.211. *Let $O(0)$ and $A(i)$ be points on the complex plane. Let point $B(z)$ passes to point C after a rotation by $90°$ around point O, point C passes to point D after a rotation by $90°$ around point A and point D passes to point B after a rotation by $90°$ around point O. What is the value of z?*

(A) 0   (B) $1+i$ (C) $-i$ (D) 1   (E) i

Solution. Answer. (E)
Note that C is the point $i \cdot z$. On the other hand $i \cdot (iz - i) + i$ corresponds to the point D. Given that
$$i \cdot (i \cdot (i \cdot z - i) + i) = z.$$

Thus, it follows that $z = i$. □

Problem 4.212. *Given numbers 1, 2, 3, 4, 5, 6, 7, 8, 9, 10. Given that Ann chooses one of those numbers, then Emily chooses one of the remaining numbers, afterward David chooses one of the remaining numbers. What is the probability that Ann's number is less than Emily's number and Emily's number is less than David's number.*

(A) $\frac{1}{2}$ (B) $\frac{2}{3}$ (C) $\frac{1}{6}$ (D) $\frac{1}{12}$ (E) $\frac{1}{24}$

Solution. Answer. (C)
Let Ann's number be a, Emily's number be b and David's number be c.
The ordered triple (a, b, c) can be chosen in $10 \cdot 9 \cdot 8$ ways.
Note that each triple can be ordered in ascending order only in one way, therefore the number of triples (m, n, p), where $m < n < p$, is equal to
$$\frac{10 \cdot 9 \cdot 8}{3!}.$$

Hence, the required probability is equal to
$$\frac{\frac{10 \cdot 9 \cdot 8}{3!}}{10 \cdot 9 \cdot 8} = \frac{1}{6}.$$

□

Problem 4.213. *Given four positive integers, such that one of them is one-digit number, one is a two-digit number, one is a three-digit number and one is a four-digit number. Given that the sum of two-digit and four-digit numbers is more by 2 than the sum of one-digit and three-digit numbers. What is the value of the sum of one-digit and three-digit numbers?*

(A) 1008   (B) 11100   (C) 900   (D) 801   (E) 716

Solution. Answer. (A)
Note that those numbers are 9, 10, 999, 1000.
Thus, it follows that $9 + 999 = 1008$. $\square$

Problem 4.214. *What is the value of the sum of all solutions of the following equation?*

$$6^x - 4 \cdot 3^x - 27 \cdot 2^x + 108 = 0.$$

(A) 1   (B) 4   (C) 5   (D) 6   (E) 2

Solution. Answer. (C)
Note that the given equation can be rewritten in the following way

$$3^x(2^x - 4) - 27(2^x - 4) = 0.$$

Thus, it follows that

$$(3^x - 27)(2^x - 4) = 0.$$

We deduce that, either

$$3^x - 27 = 0,$$

or

$$2^x - 4 = 0.$$

Therefore, either $x = 3$ or $x = 2$. Hence, the answer is $3 + 2 = 5$. $\square$

Problem 4.215. *Let a_1, a_2, a_3, a_4, a_5 be five pairwise different positive integers, such that a_1, a_2, a_3, a_4, a_5 is a geometric sequence. Given that the number of positive divisors of each of the numbers a_1 and a_5 is 5. What is the total number of positive divisors of a_4?*

(A) 5   (B) 4   (C) 6   (D) 7   (E) 8

Solution. Answer. (E)
Note that $a_1 = p^4$ and $a_5 = q^4$, where p and q are distinct primes.
The common ratio of considered geometric sequence is $\dfrac{q}{p}$. Thus, it follows that $a_4 = pq^3$. Therefore, the number of divisors of a_4 is $2 \cdot 4 = 8$. $\square$

Problem 4.216. *Let sides AD, BC of a tangential quadrilateral $ABCD$ be tangent to its incircle at points M, N, respectively. Let line segments MN and AC intersect at point E. Given that $AM = 3, CN = 6, AC = 18$. What is the value of the length of line segment AE?*

(A) 12   (B) 6   (C) 8   (D) 4.5   (E) 9

Solution. Answer. (B)
Let us choose point F on ray AD, such that $CF \parallel MN$ (see the figure).

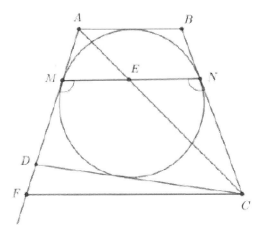

Note that
$$\angle FMN = \frac{arc(MN)}{2} = \angle CNM.$$
Therefore, quadrilateral $FMNC$ is either an isosceles trapezoid or a rectangle.
Thus, it follows that $FM = CN$.
From $\triangle FAC$ by basic proportionality theorem (Thales's theorem) we obtain that
$$\frac{AE}{EC} = \frac{AM}{MF} = \frac{AM}{CN} = \frac{1}{2}.$$
Hence, we deduce that $AE = 6$. □

Problem 4.217. *Let two cars start moving simultaneously toward each other from places A and B. Given that cars move with constant speeds. Given also that when the cars cross each other the first car coming from place A continues its movement with speed equal to 1.25 times its initial speed, and the second car continues its movement with speed equal to 1.8 times its initial speed. Afterward, it turned out that the first car reached place B at the moment when the second car reached place A. What is the ratio of the initial speed of the first car to the initial speed of the second car?*

(A) 2 (B) 1.2 (C) 2.5 (D) 3 (E) 4

Solution. Answer. (B)
Let us denote the initial speeds of the first and the second cars by V_1 and V_2, respectively. At the moment of the meeting, the first car covered the distance S_1 and the second car covered the distance S_2. Given that
$$\frac{S_1}{V_1} = \frac{S_2}{V_2},$$
and
$$\frac{S_1}{1.8V_2} = \frac{S_2}{1.25V_1}.$$
Thus, it follows that
$$\frac{V_2}{V_1} = \frac{S_2}{S_1} = \frac{1.25V_1}{1.8V_2}.$$
We deduce that
$$\left(\frac{V_1}{V_2}\right)^2 = \frac{1.8}{1.25} = \frac{36}{25}.$$
Hence, we obtain that
$$\frac{V_1}{V_2} = 1.2.$$
□

Problem 4.218. *From all positive integers not including digit 1 in their decimal notation was chosen 100th smallest number. What is the value of the sum of all digits of the chosen number?*

(A) 7   (B) 6   (C) 9   (D) 3   (E) 8

Solution. Answer. (A)
The total number of numbers less than 100 and not containing digit 1 is $9 \cdot 9 - 1 = 80$. We need to find 20th number that begins with 2 and not containing digit 1. Note that it is 232.
Hence, the answer is $2 + 3 + 2 = 7$. □

Problem 4.219. *Let M, N be points on bases BC, AD of trapezoid $ABCD$, respectively. Given that $ABMN$ and $DCMN$ are tangential quadrelaterals and $MN = 10$. What is the value of the distance between the incenters of $ABMN$ and $DCMN$?*

(A) 5   (B) 11   (C) 10   (D) 20   (E) 12

Solution. Answer. (C)
Let O_1 and O_2 be the centers of those circles (see the figure).

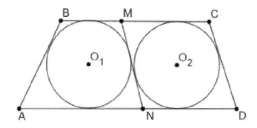

Note that
$$\angle O_1 M O_2 = \angle O_1 M N + \angle O_2 M N = \frac{1}{2}\angle BMN + \frac{1}{2}\angle CMN = 90°.$$
In a similar way, we obtain that
$$\angle O_1 N O_2 = 90°.$$
We have that
$$\angle O_1 M N + \angle O_1 N M = \frac{1}{2}\angle BMN + \frac{1}{2}\angle ANM = 90°.$$
We deduce that
$$\angle M O_1 N = 90°.$$
Thus, it follows that
$$\angle M O_2 N = 90°.$$
Therefore, $O_1 M O_2 N$ is a rectangle and its diagonals are equal. Hence
$$O_1 O_2 = MN = 10.$$

□

Problem 4.220. *The positive integer a is called "interesting number", if the value of the expression $\frac{20!}{a}$ is a perfect square. What is the number of all digits of the smallest "interesting number"?*

(A) 4   (B) 5   (C) 3   (D) 6   (E) 12

Solution. Answer. (B)
We have that
$$20! = 2^{18} \cdot 3^8 \cdot 5^4 \cdot 7^2 \cdot 11 \cdot 13 \cdot 17 \cdot 19.$$
Thus, it follows that the smallest possible "interesting number" is:
$$a = 11 \cdot 13 \cdot 17 \cdot 19 = 46189.$$
Therefore, the number of all digits of the smallest "interesting number" is equal to 5. $\square$

Problem 4.221. *Given a triangular pyramid $SABC$ with the volume V. Let V_1 be the volume of the triangular pyramid with the vertices at the intersection points of the medians of the faces of given pyramid. What is the value of $\frac{V}{V_1}$?*

(A) 27   (B) 9   (C) $\frac{9}{4}$ (D) 3   (E) 4

Solution. Answer. (A)
Let points A_1, B_1, C_1 and S_1 be the intersection points of the medians of triangles SBC, SAC, SAB, and ABC, respectively.

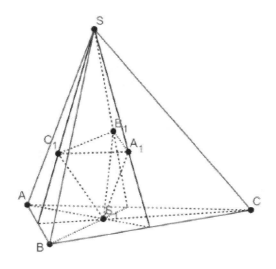

Note that $A_1C_1 \parallel AC$ and
$$A_1C_1 = \frac{2}{3} \cdot \frac{AC}{2} = \frac{AC}{3}.$$
In a similar way, we obtain that $A_1B_1 \parallel AB$ and $A_1B_1 = \frac{AB}{3}$.
Thus, it follows that
$$\frac{Area(ABC)}{Area(A_1B_1C_1)} = 9.$$
We deduce that
$$\frac{V}{V_1} = \frac{\frac{1}{3}Area(ABC) \cdot H}{\frac{1}{3}Area(A_1B_1C_1) \cdot H_1} = 9 \cdot \frac{H}{H_1} = 27.$$

Problem 4.222. What is the greatest possible value of the following expression?

$$\sin x + \sin y - \sin x \sin y.$$

(A) 3   (B) 2   (C) 1   (D) $\sqrt{3}$ (E) $\sqrt{2}$

Solution. Answer. (C)
Note that
$$\sin x + \sin y - \sin x \sin y = 1 - (1 - \sin x)(1 - \sin y) \leq 1.$$

For $x = \dfrac{\pi}{2}, y = 0$ we have that
$$\sin \frac{\pi}{2} + \sin 0 - \sin \frac{\pi}{2} \sin 0 = 1.$$

Therefore, the greatest value of the given expression is 1. □

Problem 4.223. What is the total number of all nine-digit numbers $\overline{a_1...a_9}$ with pairwise different non-zero digits, such that four numbers among the following eight two-digit numbers $\overline{a_1a_2}, \overline{a_2a_3}, ..., \overline{a_8a_9}$ are divisible by 6?

(A) 120   (B) 240   (C) 480   (D) 1680   (E) 3360

Solution. Answer. (E)
Note that in the representation of number $\overline{a_1...a_9}$ there is either 36 or 96, otherwise among the numbers $\overline{a_1a_2}, \overline{a_2a_3}, ..., \overline{a_8a_9}$ the total number of numbers divisible by 6 would not be more than 3.
Let us consider the following cases.
Case 1. If 54 is included in the representation of number $\overline{a_1...a_9}$.
If after 4 is not written either 2 or 8, then either the couple 12, 78 or the couple 72, 18 is included in the representation of number $\overline{a_1...a_8}$.
Thus, the total number of numbers $\overline{a_1...a_9}$ satisfying the assumptions of the problem is $5! \cdot 2 \cdot 2$.
If after 4 is written 2, then either 18 or 78 are included in the representation of $\overline{a_1...a_9}$. The number of all such numbers is equal to $5! \cdot 2 \cdot 2$.
If after 4 is written 8, then either 12 or 72 are included in the representation of $\overline{a_1...a_9}$. The number of all such numbers is equal to $5! \cdot 2 \cdot$.
Case 2. If 54 is not included in the representation of number $\overline{a_1...a_9}$.
Therefore, either 24 or 84 is included in the representation of number $\overline{a_1...a_9}$.
Thus, it follows that either 124 or 724 or 184 or 784 is included in the representation of number $\overline{a_1...a_9}$.
If in the representation of number $\overline{a_1...a_9}$ after 4 is written an odd digit, then the total number of numbers satisfying the assumptions of the problem is $4 \cdot 5! \cdot 2$.
If in the representation of number $\overline{a_1...a_9}$ after 4 is written an even digit, then either 1248 or 7248 or 1842 or 7842 is included in the represenation of number $\overline{a_1...a_9}$.
Hence, the total number of numbers satisfying the assumptions of the problem is $4 \cdot 5! \cdot 2$.
Thus, the answer is
$$5! \cdot 2 \cdot 2 + 4 \cdot 5! \cdot 2 + 4 \cdot 5! \cdot 2 + 8 \cdot 5! = 3360.$$

□

Problem 4.224. Let $p(x)$ and $q(x)$ be polynomials with integer coefficients, such that
$$p^2(x) - p(x)q(x) + q^2(x) = x^2 - x + 1$$

holds true for any x. What is the total number of all such pairs $(p(x), q(x))$?

(A) 0   (B) 12   (C) 8   (D) 10   (E) 16

Solution. Answer. (B)
Note that if the pair $(p(x), q(x))$ satisfies to the conditions of the problem, then the pair $(q(x), p(x))$ also satisfies the conditions of the problem.
Therefore, without loss of generality we can assume that $\deg p(x) \geq \deg q(x)$.
Note that
$$\deg(p^2(x) - p(x)q(x) + q^2(x)) = 2\deg p(x).$$
Thus, it follows that $\deg p(x) = 1$.
We obtain that
$$p(x) = mx + n, q(x) = ux + v.$$
In this case, we have that
$$m^2 - mu + u^2 = 1.$$
Hence, we deduce that
$$3m^2 + (m - 2u)^2 = 4.$$
Therefore, either $m = 1$ or $m = -1$.
In a similar way, we obtain that
$$n^2 - nv + v^2 = 1.$$
Thus, it follows that $n \in \{-1, 0, 1\}$.
Note that if the pair $(p(x), q(x))$ satisfies the conditions of the problem, then $(-p(x), -q(x))$ also satisfies the conditions of the problem.
Therefore, we can assume that either $p(x) = x - 1$ or $p(x) = x$ or $p(x) = x + 1$.
If $p(x) = x - 1$, then $q(x) = x$ or $q(x) = -1$.
If $p(x) = x$, then $q(x) = x - 1$ or $q(x) = 1$.
If $p(x) = x + 1$, then $q(x)$ is not a polynomial.
Hence, all possible pairs are the following pairs
$$(x-1, x), (x, x-1), (-x+1, -x), (-x, -x+1), (x-1, -1), (-1, x-1),$$
$$(-x+1, 1), (1, -x+1), (x, 1), (1, x), (-x, -1), (-1, -x).$$
Therefore, the correct answer is 12. $\square$

Problem 4.225. *Let a two-digit number be chosen randomly. What is the probability that the chosen number can be represented as a difference of cubes of two positive integers?*

(A) $\dfrac{1}{2}$ (B) $\dfrac{1}{10}$ (C) $\dfrac{4}{45}$ (D) $\dfrac{1}{15}$ (E) $\dfrac{2}{3}$

Solution. Answer. (C)
Assume that two-digit number $\overline{ab}$ can be represented as $n^3 - m^3$, where m and n are positive integers.
Note that $n \geq m + 1$. Thus, it follows that
$$99 \geq \overline{ab} \geq (m+1)^3 - m^3 = 3m^2 + 3m + 1.$$
Hence, we obtain that $m \in \{1, 2, 3, 4, 5\}$.
If $m = 1$, then in order $n^3 - m^3$ to be a two-digit number we should have $n = 3$ or $n = 4$. Therefore $\overline{ab} = 26$ or $\overline{ab} = 63$.
If $m = 2$, then $n = 3$ or $n = 4$. Therefore $\overline{ab} = 19$ or $\overline{ab} = 56$.
If $m = 3$, then $n = 4$ or $n = 5$. Therefore $\overline{ab} = 37$ or $\overline{ab} = 98$.
If $m = 4$, then $n = 5$. Therefore $\overline{ab} = 61$.
If $m = 5$, then $n = 6$. Therefore $\overline{ab} = 91$.
Therefore, the required probability is
$$\frac{8}{9 \cdot 10} = \frac{4}{45}.$$
$\square$

4.10 Solutions of AMC 12 type practice test 10

Problem 4.226. *What is the value of the following expression?*

$$\frac{(-1)^{-1}}{(-3)^{-3}} + \frac{1}{(-2)^{-2}}.$$

(A) 31 (B) 23 (C) -23 (D) $\frac{31}{108}$ (E) -31

Solution. Answer. (A)

$$\frac{(-1)^{-1}}{(-3)^{-3}} + \frac{1}{(-2)^{-2}} = \frac{-1}{-\frac{1}{27}} + (-2)^2 = 27 + 4 = 31.$$

□

Problem 4.227. *James is repairing torn pages in each of 4 books. He spends the same amount of time for the reparation of each book and after the reparation of each book he rests for 3 minutes. Given that James started to fix the first book at 14 : 00 and finished the third book at 15 : 06. At what time was James done with reparation of all 4 books?*

(A) 15 : 28 (B) 15 : 26 (C) 15 : 29 (D) 15 : 30 (E) 16 : 00

Solution. Answer. (C)
Note that the reparation of each book lasted $(66 - 6) : 3 = 20$ minutes.
Therefore, he finished the work at 15 : 29.

□

Problem 4.228. *An archer shot several arrows and each time he got either 5 or 7 points. All together, the archer received 58 points. What is the total number of shots?*

(A) 11 (B) 10 (C) 12 (D) 8 (E) 6

Solution. Answer. (B)
Let us assume that the gunman got 5 points x times and 7 points y times. Given that

$$5x + 7y = 58.$$

Thus, it follows that

$$5 \mid (58 - 7y).$$

Hence, we obtain that $y = 4, x = 6$. Therefore, the answer is $6 + 4 = 10$.

□

Problem 4.229. *Jack, Marta, Rand, Todd and 6 more students went to pick mushrooms. Every two of these 10 pupils picked a different number of mushrooms and all together they picked 45 mushrooms. Jack and Marta together picked 16, Marta and Rand together picked 15, Rand and Todd together picked 13 mushrooms. How many mushrooms did Todd pick?*

(A) 10 (B) 7 (C) 5 (D) 6 (E) 0

Solution. Answer. (C)
Note that those 10 pupils picked 0, 1, 2,..., 9 mushrooms.
If Jack picked 9 mushrooms, then Marta picked 7, Rand picked 8 and Todd picked 5 mushrooms.
If jack picked 7 mushrooms, then we obtain that Todd also picked 7. This leads to a contradiction, as it is given that all students picked different number of mushrooms.
Therefore, Todd picked 5 mushrooms.

□

Problem 4.230. *A tourist group consists of French and Chinese tourists, such that none of the group members has dual citizenship. Given that 15 % of French tourists and 10% of Chinese tourists (in this group) speak english. What is the smallest possible number of tourists in the group?*

(A) 15 (B) 35 (C) 20 (D) 25 (E) 30

Solution. Answer. (E)
Let the number of French tourists be x and the number of Chinese tourists be y. Given that
$$\frac{x \cdot 15}{100} = \frac{3x}{20},$$
and
$$\frac{y \cdot 10}{100} = \frac{y}{10},$$
are positive integers. Thus, it follow that $x \geq 20$ and $y \geq 10$.
Therefore, the smallest possible value of $x + y$ is equal to 30.
Hence, we obtain that the smallest possible number of tourists in this group is equal to 30. □

Problem 4.231. *Let the columns of 10×10 square grid be numbered from left to right by numbers from 1 to 10, and the rows be numbered from bottom to top by numbers from 1 to 10. Let the intersection point of ith column and j^{th} row be number $ij + i + j$. What percent of all numbers in the square are even numbers?*

(A) 75 (B) 50 (C) 30 (D) 25 (E) 35

Solution. Answer. (D)
We have that
$$ij + i + j = (i+1)(j+1) - 1.$$
From this representation we easily note that $ij + i + j$ is even, when i and j are both even.
As there are 5 possible even values for i and 5 possible even values for j, thus the total number of even numbers of the form $ij + i + j$ is equal to $5 \cdot 5$. Note that 10×10 square grid in total contains 100 numbers, hence 25 percent of all numbers in the square are even numbers. □

Problem 4.232. *The regular $n-$gon is divided into several regular $m-$gons, where $n > m$. What is the value of $m + n$?*

(A) 9 (B) 6 (C) 7 (D) 8 (E) 10

Solution. Answer. (A)
We have that
$$\frac{180°(n-2)}{n} = \frac{180°(m-2)}{m} \cdot k,$$
where k is a positive integer. Note that
$$n(m-2) \mid m(n-2).$$
Thus, it follows that
$$n(m-2) \mid (m(n-2) - n(m-2)).$$
Hence, we obtain that
$$2(n-m) \geq n(m-2).$$
We have that $m = 3$.
In this case, we deduce that
$$n \mid 3(n-2).$$
Therefore $n = 6$. Thus, it follows that $n + m = 9$. □

Problem 4.233. *What is the value of the following expression*

$$\frac{2^{\log_2^2 3}}{3^{\log_2 12}}.$$

(A) 1   (B) $\frac{1}{9}$ (C) 9   (D) 3   (E) 2

Solution. Answer. (B)
Let $\log_2 3 = a$. Then, we have that

$$\log_2 12 = \log_2(3 \cdot 4) = 2 + \log_2 3 = 2 + a.$$

On the other hand, we have that $2^a = 3$. Therefore

$$\frac{2^{a^2}}{3^{2+a}} = \frac{3^a}{9 \cdot 3^a} = \frac{1}{9}.$$

$\square$

Problem 4.234. *There are 3 black and 4 white balls in a bag. Mary and Ann one after the other each took a single ball. Mary starts and the girls look at the color of the ball after taking it. Whoever takes a white ball first is the winner. What is the probability that Ann wins?*

(A) $\frac{24}{35}$ (B) $\frac{1}{2}$ (C) $\frac{11}{35}$ (D) $\frac{1}{3}$ (E) $\frac{2}{3}$

Solution. Answer. (C)
Let us denote by B a black ball and by W a white ball. Note that Ann wins if she takes in the pattern BW or $BBBW$. Therefore, the probability that Ann wins is

$$\frac{3}{7} \cdot \frac{4}{6} + \frac{3}{7} \cdot \frac{2}{6} \cdot \frac{1}{5} = \frac{11}{35}.$$

$\square$

Problem 4.235. *Let m and n be the number of all pairwise non-congruent triangles with integer sides, such that each of them has a perimeter of 2017 and 2018, respectively. What is the value of $n - m$?*

(A) 0   (B) 168   (C) -168   (D) 5   (E) -5

Solution. Answer. (C)
Let M be the set of all integer triples (a, b, c), such that

$$\begin{cases} 1 \le a \le b \le c, \\ a + b > c, \\ a + b + c = 2017. \end{cases}$$

Let N be the set of all integer triples (a, b, c), such that

$$\begin{cases} 1 \le a \le b \le c, \\ a + b > c, \\ a + b + c = 2018. \end{cases}$$

Given that the number of elements of sets M and N are m and n, respectively.
For $(a, b, c) \in M$, we have that $(a, b, c+1) \notin N$ when $a + b = c + 1$.
Note that the number of such triples is equal to 504, because $c = 1008$ and $a + b = 1009$.

For $(a, b, c) \in N$, we have that $(a, b, c - 1) \notin M$ when $b = c$.
Note that the number of such triples is equal to 336, because $a + 2b = 2018$.
Thus, it follows that
$$n - m = 336 - 504 = -168.$$

$\square$

Problem 4.236. *What is the area of a triangle with sides lying on lines given by the following equations $x = 4$, $y = 3$ and $x + y = 11$?*

(A) 16   (B) 28   (C) 14   (D) 20   (E) 8

Solution. Answer. (E)
Note that these lines pairwisely intersect at points $C(4, 3)$, $A(4, 7)$, $B(8, 3)$. Therefore, the required triangle is a right triangle ABC with legs of length of 4 and 4. Hence, its area is $\dfrac{4 \cdot 4}{2} = 8$. $\square$

Problem 4.237. *Given that pairwise different numbers a, b, c, d form an arithmetic sequence and their sum is equal to 20. What is the sum of all solutions of the following equation?*
$$(x - a)(x - b)(x - c) + (x - b)(x - c)(x - d) = 0.$$

(A) 20   (B) 10   (C) 12   (D) 15   (E) 40

Solution. Answer. (D)
Note that
$$0 = (x - a)(x - b)(x - c) + (x - b)(x - c)(x - d) = (x - b)(x - c)(2x - a - d).$$

Therefore, the solutions of given equation are numbers $b, c, \dfrac{a + d}{2}$.
As a, b, c, d form an arithmetic sequence and their sum is equal to 20, then we have that
$$a + d = b + c = 10.$$

Thus, it follows that the sum of all solutions of given equation is:
$$b + c + \dfrac{a + d}{2} = 10 + \dfrac{10}{2} = 15.$$

$\square$

Problem 4.238. *Let $ABCD$ be a cyclic quadrilateral, where $\angle BAC + \angle ADB = 60°$ and $AB = 7$, $BC = 8$. What is the value of the length of AC?*

(A) 8   (B) 10   (C) 11   (D) 13   (E) 12

Solution. Answer. (D)
Note that $\angle ADB = \angle ACB$ (see the figure).
Thus, it follows that
$$\angle ABC = 180° - (\angle BAC + \angle ACB) = 180° - (\angle BAC + \angle ADB) = 120°.$$

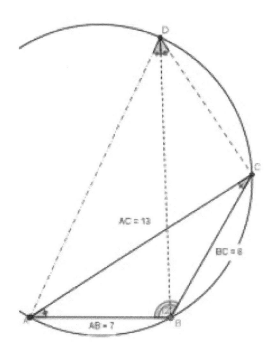

According to the law of cosines, we have that

$$AC^2 = AB^2 + BC^2 - 2 \cdot AB \cdot BC \cdot \cos 120° = 7^2 + 8^2 + 7 \cdot 8 = 169.$$

Hence, we obtain that $AC = 13$. □

Problem 4.239. *Given an equilateral triangle with side length of 6 and a circle intersecting this triangle, such that its diameter is one of the sides of the triangle. What is the area of the part of the circle that is inside of the triangle?*

(A) $\dfrac{3(\pi + 3\sqrt{3})}{2}$ (B) $\dfrac{27\sqrt{3}}{4}$ (C) $\dfrac{9\pi}{2}$ (D) $\dfrac{3(2\pi + 3\sqrt{3})}{2}$ (E) $\dfrac{27}{2}$

Solution. Answer. (A) The part with required area consists of two equilateral triangles with side length of 3 and a circular sector of radius 3 and a central angle 60° (see the figure).

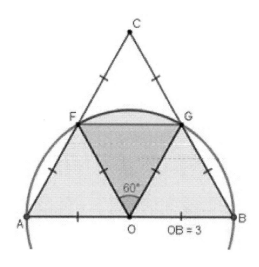

Therefore, the required area is
$$2 \cdot \frac{\sqrt{3}}{4} \cdot 3^2 + \frac{\pi \cdot 3^2}{360} \cdot 60 = \frac{3(\pi + 3\sqrt{3})}{2}.$$

□

Problem 4.240. *Let 1×1 cell be randomly removed from 7×7 square grid. What is the probability that (after 1×1 cell was removed) the remaining part of 7×7 square grid can be covered by sixteen 1×3 rectangular grids?*

(A) $\dfrac{1}{2}$ (B) $\dfrac{9}{49}$ (C) $\dfrac{8}{49}$ (D) $\dfrac{16}{49}$ (E) $\dfrac{3}{4}$

Solution. Answer. (B)

Let us write numbers 1, 2, 3 in the cells of given 7×7 square grid, such that in each cell is written one single number (see the figure).

1	2	3	1	2	3	1
3	1	2	3	1	2	3
2	3	1	2	3	1	2
1	2	3	1	2	3	1
3	1	2	3	1	2	3
2	3	1	2	3	1	2
1	2	3	1	2	3	1

Note that the total number of 1's, 2's and 3's are 17, 16, 16, respectively.

Note also that in each 1×3 rectangular grid there are numbers 1, 2, 3.

Therefore, in order to cover the remaining part of 7×7 square grid by sixteen 1×3 rectangular grids, we need to remove a cell with 1 written in it (in this case the total number of 1's will be equal to 16 also).

1	3	2	1	3	2	1
3	2	1	3	2	1	3
2	1	3	2	1	3	2
1	3	2	1	3	2	1
3	2	1	3	2	1	3
2	1	3	2	1	3	2
1	3	2	1	3	2	1

Therefore, we need to remove one of 9 cells with X (see the figure).

X			X			X
X			X			X
X			X			X

After removing any of those 9 cells, the remaining part can be divided into rectangular grids, such that one of the dimensions of each of those rectangles is a multiple of 3.

Hence, the probability that (after 1×1 cell was removed) the remaining part of 7×7 square grid can be covered by sixteen 1×3 rectangular grids is equal to $\frac{9}{49}$. □

Problem 4.241. *What is the value of the volume of a quadrilateral pyramid, such that its all eight edges have a length of 6?*

(A) 72 (B) $72\sqrt{2}$ (C) $36\sqrt{2}$ (D) 36 (E) $48\sqrt{2}$

Solution. Answer. (C)

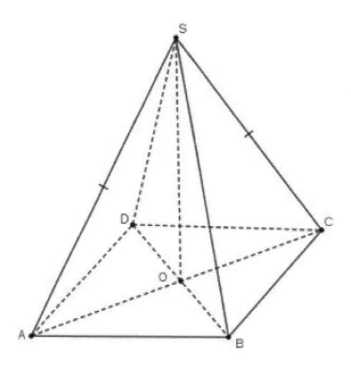

Let the quadrangular pyramid be $SABCD$. Given that $ABCD$ is a rhombus. Let O be the intersection point of diagonals of $ABCD$. Note that $SA = SC$. Thus, it follows that $SO \perp AC$.
In a similar way, we can obtain that $SO \perp BD$.
As triangles SOA and SOB are congruent, we deduce that $OA = OB$. Therefore $ABCD$ is a square. We also obtain that SO is the height of the pyramid.
Given that
$$OA = \frac{AC}{2} = 3\sqrt{2},$$
and
$$SO = \sqrt{SA^2 - OA^2} = \sqrt{36 - 18} = 3\sqrt{2}.$$
Thus, it follows that
$$Volume(SABCD) = \frac{1}{3} \cdot 36 \cdot 3\sqrt{2} = 36\sqrt{2}.$$

□

Problem 4.242. Jack takes without looking n balls out of a box that contains white, red and blue balls. The probability of taking out a white ball is $\frac{1}{2}$, the probabilities of taking out red and blue balls are $\frac{1}{3}$ and $\frac{1}{6}$, respectively. Given that the probability that "out of n balls exactly 2 balls are white and exactly 3 balls are red" is equal to the probability that "out of n balls exactly 3 balls are white and exactly 3 balls are red." What is the value of n?

(A) 10   (B) 9   (C) 8   (D) 7   (E) 6

Solution. Answer. (E)
Note that probability of the event that out of n balls exactly 2 balls are white and exactly 3 balls are red is
$$\binom{n}{2}\left(\frac{1}{2}\right)^2\binom{n-2}{3}\left(\frac{1}{3}\right)^3\left(\frac{1}{6}\right)^{n-5}.$$
On the other hand, probability of the event that out of n balls exactly 3 balls are white and exactly 3 balls are red is
$$\binom{n}{3}\left(\frac{1}{2}\right)^3\binom{n-3}{3}\left(\frac{1}{3}\right)^3\left(\frac{1}{6}\right)^{n-6}.$$
Thus, it follows that
$$\frac{n(n-1)}{2}\cdot\frac{(n-2)(n-3)(n-4)}{6}\cdot\frac{1}{6}=\frac{n(n-1)(n-2)}{6}\cdot\frac{(n-3)(n-4)(n-5)}{6}\cdot\frac{1}{2}.$$
Hence, we obtain that $n - 5 = 1$. Therefore $n = 6$. □

Problem 4.243. A positive integer is called a "special number" if it has at least three different prime divisors. Let n be a "special number", we denote by $Q(n)$ the sum of the pairwise products of all distinct prime divisors of n, for example
$$Q(30) = 2\cdot 3 + 3\cdot 5 + 2\cdot 5 = 31,$$
$$Q(60) = 2\cdot 3 + 3\cdot 5 + 2\cdot 5 = 31.$$
What is the smallest possible even value of $Q(n)$?

(A) 236   (B) 32   (C) 64   (D) 100   (E) 320

Solution. Answer. (A)
Note that if n has three prime divisors, then $Q(n)$ is odd.
If n has four prime divisors and $Q(n)$ is even, then n is odd.
Therefore, the answer is
$$3\cdot 5 + 3\cdot 7 + 3\cdot 11 + 5\cdot 7 + 5\cdot 11 + 7\cdot 11 = 236.$$
□

Problem 4.244. Let ABC be a right triangle, such that $\angle C = 90°$. Given that regular hexagons $AC_1C_2C_3C_4C$ and $AB_1B_2B_3B_4B$ are constructed externally on the sides of triangle ABC, such that points C_2, C_3, B_2, B_3 lie on one circle. What is the value (in degrees) of $\angle B$?

(A) 30   (B) 45   (C) 15   (D) 60   (E) 75

Solution. Answer. (D)

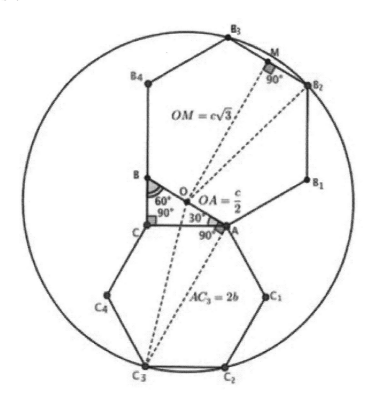

Let $AC = b$ and $AB = c$. Note that the perpendicular bisector of line segment C_2C_3 passes through point O, where O is the midpoint of side AB. On the other hand, the perpendicular bisector of segment B_2B_3 coincides with the perpendicular bisector of line segment AC, therefore it also passes through point O. Given that $OC_3 = OB_2$.

From right triangles OAC_3 and OMB_2 according to the Pythagorean theorem, we obtain that

$$OC_3^2 = OA^2 + AC_3^2 = 4b^2 + \frac{c^2}{4},$$

and

$$OB_2^2 = OM^2 + MB_2^2 = \frac{13c^2}{4},$$

where M is the midpoint of line segment B_2B_3. We obtain that

$$\frac{13c^2}{4} = 4b^2 + \frac{c^2}{4}.$$

Thus, it follows that $b = \frac{\sqrt{3}}{2}c$. Therefore $\angle B = 60°$. □

Problem 4.245. Let function f be defined as follows $f : \{0, 1, 2, 3, ...\} \times \{0, 1, 2, 3, ...\} \to \{0, 1, 2, 3, ...\}$, $f(0,0) = 2$ and

$$f(i,j) = \begin{cases} 2f(i-1, j) - 1, & \text{if } j = 0, i \geq 1, \\ 3f(i, j-1) - 2, & \text{if } i = 0, j \geq 1, \\ f(i-1, j) + f(i, j-1) - f(i-1, j-1), & \text{if } i \geq 1, j \geq 1. \end{cases}$$

What is the value of $f(3,4)$?

(A) 7 (B) 89 (C) 25 (D) 91 (E) 100

Solution. Answer. (B)
Let us consider the following table, where in the left bottom corner is written $f(0,0) = 2$, the rest of the numbers written in the first row we find using the following formula

$$f(i, 0) = 2f(i-1, 0) - 1,$$

for $i = 1, 2, 3$, the rest of the numbers written in the first column we find using the following formula

$$f(0, j) = 3f(0, j-1) - 2,$$

for $j = 1, 2, 3, 4$.

4	82	83	85	89
3	28	29	31	35
2	10	11	13	17
1	4	5	7	11
0	2	3	5	9
	0	1	2	3

Note that the numbers written in all other entries we find using 2×2 squares provided below, starting from the left bottom corner and writting in the empty square $m + n - k$.

m	
k	n

We write $m + n - k$, because we use the following formula

$$f(i, j) = f(i-1, j) + f(i, j-1) - f(i-1, j-1),$$

for $i = 1, 2, 3$ and $j = 1, 2, 3, 4$.
Thus, it follows that $f(3, 4) = 89$.
Alternative solution. Applying mathematical induction with respect to $i + j$ one can easily prove that

$$f(i, j) = 2^i + 3^j.$$

Thus, it follows that $f(3, 4) = 2^3 + 3^4 = 89$. □

Problem 4.246. *A three-digit number is called "ordinary", if the positive difference of its some two neighbor digits is greater than or equal to 2. What is the total number of "ordinary" three-digit numbers?*

(A) 823   (B) 825   (C) 650   (D) 640   (E) 700

Solution. Answer. (B)

At first, let us find the total number of all not "ordinary" three-digit numbers.

Let $\overline{abc}$ be not an "ordinary" three-digit number. Thus, it follows that

$$|a - b| \leq 1,$$

and

$$|b - c| \leq 1.$$

Hence, we deduce that

$$|a - c| \leq 2.$$

Let us consider the following cases.

Case 1. If $a - c = 2$.
Case 2. If $a - c = -2$.
Case 3. If $a - c = 1$.
Case 4. If $a - c = -1$.
Case 5. If $a - c = 0$.

Note that the total number of three-digit numbers satisfying to case 1 and to conditions $|a - b| \leq 1$, $|b - c| \leq 1$ is equal to 8.

The total number of three-digit numbers satisfying to case 2 and to conditions $|a - b| \leq 1$, $|b - c| \leq 1$ is equal to 7.

The total number of three-digit numbers satisfying to case 3 and to conditions $|a - b| \leq 1$, $|b - c| \leq 1$ is equal to 18.

The total number of three-digit numbers satisfying to case 4 and to conditions $|a - b| \leq 1$, $|b - c| \leq 1$ is equal to 16.

The total number of three-digit numbers satisfying to case 5 and to conditions $|a - b| \leq 1$, $|b - c| \leq 1$ is equal to 26.

Therefore, the total number of three-digit "ordinary" numbers is equal to 825, as $900 - 75 = 825$. □

Problem 4.247. *In a meeting 32 chairs are placed around a circular table and each of them is occupied by one person. After a break, each of them can occupy the eighth chair (in any direction) counting from the chair they have occupied before the break, where the counting starts from an adjacent chair (after the break also each chair is occupied by one person). After the break, in how many ways can these people be seated around the table?*

(A) 65536   (B) 128   (C) 32   (D) 2^{32} (E) 16

Solution. Answer. (A)

Let us enumerate the chairs (in some direction) with numbers 1, 2, 3,..., 32.

Note that people who occupied the chairs with numbers 1, 9, 17, 25 again occupy these four chairs and they do that exactly in four ways.

Note also that we can divide 32 chairs into 8 such groups.

Thus, it follows that after the break these people can be seated around the table in 65536 ways, as $4^8 = 65536$. □

Problem 4.248. *Given a rectangular box with dimensions $a \times b \times c$, where a, b, c are positive integers and $a \leq b \leq c$. The volume and the sum of the lengths of all edges of this rectangular box are numerically equal. How many such ordered triples (a, b, c) are possible?*

(A) 10 (B) 6 (C) 14 (D) 11 (E) 8

Solution. Answer. (B)
As the volume and the sum of the lengths of all edges of given rectangular box are numerically equal, then we obtain that
$$abc = 4(a + b + c).$$
Given that $a \leq b \leq c$, hence
$$4(a + b + c) \leq 4 \cdot 3c.$$
As $c > 0$, we deduce that
$$ab \leq \frac{4(a + b + c)}{c} \leq 12.$$
On the other hand, as $a \leq b$, we have that
$$a^2 \leq ab \leq 12.$$
Hence, as a is a positive integer, then either $a = 1$ or $a = 2$ or $a = 3$.
Let us consider the following cases.
Case 1. If $a = 1$, then
$$(b - 4)(c - 4) = 20.$$
Thus, it follows that
$$b - 4 = 1, \ c - 4 = 20,$$
or
$$b - 4 = 2, \ c - 4 = 10,$$
or
$$b - 4 = 4, \ c - 4 = 5.$$
Thus, in this case we obtain that either $(a, b, c) = (1, 5, 24)$ or $(a, b, c) = (1, 6, 14)$ or $(a, b, c) = (1, 8, 9)$.
Case 2. If $a = 2$, then we obtain that
$$(b - 2)(c - 2) = 8.$$
We deduce that, either
$$b - 2 = 1, \ c - 2 = 8,$$
or
$$b - 2 = 2, \ c - 2 = 4.$$
Thus, in this case we obtain that either $(a, b, c) = (2, 3, 10)$ or $(a, b, c) = (2, 4, 6)$.
Case 3. If $a = 3$, then we obtain that
$$(3b - 4)(3c - 4) = 52.$$
Thus, it follows that
$$3b - 4 = 2, \ 3c - 4 = 26.$$
Thus, in this case we obtain that $(a, b, c) = (3, 2, 10)$.
Therefore, in total there are 6 such ordered triples. $\square$

Problem 4.249. *Let ABC be an obtuse non-isosceles triangle. Let M and N be points on the longest side AC, such that $AM = CN = 39$ and $AB : BM = BC : BN = 8 : 5$. Given that the lenght of the altitude of triangle ABC drawn to side AC is 24. What is the length of AC?*

(A) 40   (B) 36   (C) 100   (D) 64   (E) 128

Solution. Answer. (E)

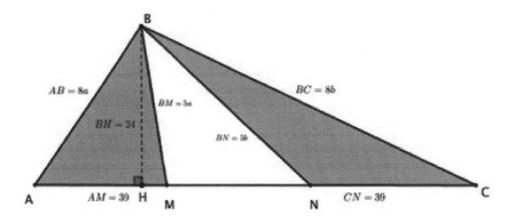

Let $AB < BC$ and $AB = 8a$, $BC = 8b$. Then $BM = 5a$ and $BN = 5b$. As $BH = 24$, then

$$Area(\triangle ABM) = Area(\triangle BCN) = 39 \cdot 12.$$

By Heron's formula, we have that

$$39 \cdot 12 = \sqrt{\frac{13a+39}{2} \cdot \frac{13a-39}{2} \cdot \frac{39-3a}{2} \cdot \frac{39+3a}{2}} = \sqrt{\frac{13b+39}{2} \cdot \frac{13b-39}{2} \cdot \frac{39-3b}{2} \cdot \frac{39+3b}{2}}.$$

Hence, we deduce that

$$39 \cdot 12 = \frac{13 \cdot 3}{4}\sqrt{(a+3)(a-3)(13-a)(13+a)} = \frac{13 \cdot 3}{4}\sqrt{(b+3)(b-3)(13-b)(13+b)}.$$

Using the identity $(x-y)(x+y) = x^2 - y^2$ we obtain that

$$48 = \sqrt{(a^2 - 3^2)(13^2 - a^2)} = \sqrt{(b^2 - 3^2)(13^2 - b^2)}.$$

Therefore a^2 and b^2 are the solutions of the following equation

$$48 = \sqrt{(x-9)(169-x)}.$$

Thus, it follows that

$$a^2 = 25,$$

and

$$b^2 = 153.$$

Applying the Pythagorean theorem in right triangles ABH and BHC, we obtain that

$$AC = AH + HC = \sqrt{AB^2 - BH^2} + \sqrt{BC^2 - BH^2} = \sqrt{64 \cdot 25 - 24^2} + \sqrt{64 \cdot 153 - 24^2} = 128.$$

□

Problem 4.250. *Let polynomial $p(x)$ be given as follows:*
$$p(x) = 1 + 2x + 3x^2 + \ldots + 2015x^{2014}.$$
What is the value of $\left|p\left(e^{i\frac{\pi}{3}}\right)\right|?$

(A) 0   (B) 2014   (C) 2016   (D) 1   (E) 2015

Solution. Answer. (C)

Let
$$z = e^{i\frac{\pi}{3}}.$$
Taking this into consideration and using Euler's formula, we obtain that
$$z^6 = e^{i \cdot 2\pi} = \cos 2\pi + i \sin 2\pi = 1 + 0 = 1.$$
We have that
$$p(z) = 1 + 2z + 3z^2 + \ldots + 2015z^{2014}.$$
Multiplying both sides of this equation by z, we obtain that
$$zp(z) = z + 2z^2 + 3z^3 + \ldots + 2015z^{2015}.$$
Subtracting the last equation from the previous one, we deduce that
$$p(z) - zp(z) = 1 + z + z^2 + \ldots + z^{2014} - 2015\frac{(z^6)^{336}}{z} =$$
$$= \frac{z^{2015} - 1}{z - 1} - \frac{2015}{z} = \frac{\frac{1}{z} - 1}{z - 1} - \frac{2015}{z} = -\frac{2016}{z}.$$
Note that $z \neq 1$, as using Euler's formula, we obtain that
$$z = e^{i\frac{\pi}{3}} = \cos\frac{\pi}{3} + i \sin\frac{\pi}{3} = \frac{1}{2} + \frac{\sqrt{3}}{2}i \neq 1.$$
Therefore we are allowed to divide $p(z) - zp(z)$ by $z - 1$, thus
$$p(z) = \frac{2016}{z(z-1)}.$$
Hence, we obtain that
$$|p(z)| = \frac{2016}{|z(z-1)|} = \frac{2016}{|z| \cdot |z-1|} = \frac{2016}{1 \cdot 1} = 2016,$$
as
$$|z| = \left|\frac{1}{2} + \frac{\sqrt{3}}{2}i\right| = \sqrt{\left(\frac{1}{2}\right)^2 + \left(\frac{\sqrt{3}}{2}\right)^2} = 1,$$
and
$$|z - 1| = \left|-\frac{1}{2} + \frac{\sqrt{3}}{2}i\right| = \sqrt{\left(-\frac{1}{2}\right)^2 + \left(\frac{\sqrt{3}}{2}\right)^2} = 1.$$
Thus, it follows that
$$\left|p\left(e^{i\frac{\pi}{3}}\right)\right| = 2016.$$

□

4.11 Solutions of AMC 12 type practice test 11

Problem 4.251. *What is the value of the following expression*

$$\sqrt[6]{2^5 \cdot \sqrt[7]{2^6 \cdot \sqrt[8]{2^7 \cdot \sqrt[9]{2^8 \cdot \sqrt[10]{2^{10}}}}}}.$$

(A) $\sqrt[6]{2}$ (B) 2 (C) 4 (D) 8 (E) 1024

Solution. Answer. (B)

$$\sqrt[6]{2^5 \cdot \sqrt[7]{2^6 \cdot \sqrt[8]{2^7 \cdot \sqrt[9]{2^8 \cdot \sqrt[10]{2^{10}}}}}} = \sqrt[6]{2^5 \cdot \sqrt[7]{2^6 \cdot \sqrt[8]{2^7 \cdot \sqrt[9]{2^8 \cdot 2}}}} =$$

$$= \sqrt[6]{2^5 \cdot \sqrt[7]{2^6 \cdot \sqrt[8]{2^7 \cdot 2}}} = \sqrt[6]{2^5 \cdot \sqrt[7]{2^6 \cdot 2}} = \sqrt[6]{2^5 \cdot 2} = 2.$$

□

Problem 4.252. *Let the initial speed of a car be 70 miles per hour. At first its speed was increased by 10%, then the obtained speed was decreased by 10 miles per hour. What is the value (in miles per hour) of the final speed of the car?*

(A) 70 (B) 69 (C) 68 (D) 67 (E) 50

Solution. Answer. (D)
Note that after increasing the speed of the car by 10% it's speed will be equal to the folliwng number (in miles per hour)

$$70 + \frac{70 \cdot 10}{100} = 77.$$

Therefore, the value (in miles per hour) of the final speed of the car is $77 - 10 = 67$. □

Problem 4.253. *What is the value of the product of all solutions of the following equation?*

$$x\left(x^2 - 4 + \frac{1}{x}\right) = 1.$$

(A) 0 (B) -4 (C) 4 (D) 1 (E) -1

Solution. Answer. (B)
Note that the given equation is equivalent to the following system

$$\begin{cases} x(x^2 - 4) = 0, \\ x \neq 0. \end{cases}$$

Therefore, the solutions of the given equation are numbers 2 and -2.
Thus, it follows that their product is equal to $(-2) \cdot 2 = -4$. □

Problem 4.254. *Let a be a positive integer, such that the difference of the arithmetic mean and geometric mean of numbers a and 9 is equal to 8. What is the value of the sum of all digits of a?*

(A) 13   (B) 14   (C) 15   (D) 20   (E) 21

Solution. Answer. (A)
Given that
$$\frac{a+9}{2} - \sqrt{a \cdot 9} = 8.$$
Thus, it follows that
$$(\sqrt{a} - 3)^2 = 16.$$
We deduce that
$$\sqrt{a} - 3 = 4,$$
as the case $\sqrt{a} - 3 = -4$ is not possible.
Therefore $a = 49$. Hence, the sum of all digits of a is $4 + 9 = 13$. □

Problem 4.255. *Let CD be the bisector of angle ACB in triangle ABC. Given that $AC = CD$ and $\angle ACB = 108°$. Let $\angle A = n \cdot \angle B$. What is the value of n?*

(A) 3   (B) 5   (C) 6   (D) 7   (E) 10

Solution. Answer. (D)
We have that
$$\angle ACD = \frac{\angle C}{2} = 54°,$$
and
$$\angle CAD = \angle CDA = \frac{180° - 54°}{2} = 63°.$$
Thus, it follows that $\angle A = 63°$ and
$$\angle B = 180° - (108° + 63°) = 9°.$$
Therefore $n = 7$. □

Problem 4.256. *As a homework a student needs to solve algebra and geometry problems, all together 27 problems. Given that the student has managed to solve 80% of algebra problems and 75% of geometry problems. What part of the entire homework has the student finished?*

(A) $\dfrac{7}{9}$ (B) $\dfrac{2}{3}$ (C) $\dfrac{5}{9}$ (D) $\dfrac{1}{2}$ (E) $\dfrac{3}{7}$

Solution. Answer. (A)
According to the condition of the problem the number of algebra problems is a multiple of 5 and the number of geometry problems is a multiple of 4. On the other hand, given that their sum is equal to 27. Therefore, straightforward verification shows that the number of algebra problems is equal to 15 and the number of geometry problems is equal to 12.
Hence, the student has managed to solve 21 problems (as $12 + 9 = 21$).
Thus, it follows that the student has finished the following part of the homework:
$$\frac{21}{27} = \frac{7}{9}.$$
□

Problem 4.257. *What is the total number of all integers a, such that the following equation has exactly two real solutions?*
$$|2^x - 10| = a.$$

(A) 21   (B) 19   (C) 11   (D) 10   (E) 9

Solution. Answer. (E)
From the given equation we obtain that, either
$$\begin{cases} 2^x = a + 10, \\ a \geq 0, \end{cases}$$
or
$$\begin{cases} 2^x = 10 - a, \\ a \geq 0. \end{cases}$$

Therefore, the given equation has exactly two solutions if
$$\begin{cases} a > 0, \\ a + 10 > 0, \\ 10 - a > 0. \end{cases}$$

Thus, it follows that $a \in (0, 10)$. Hence, there are 9 integer values of a. □

Problem 4.258. *Let in all cells of 3×3 square grid be written positive integers (one number per cell), such that the sum of all written numbers is equal to 215 and their product is equal to 2020. How many 1 is written in the cells of 3×3 square grid?*

(A) 3   (B) 4   (C) 5   (D) 6   (E) 7

Solution. Answer. (D)
We have that
$$2020 = 2^2 \cdot 5 \cdot 101.$$

Therefore, in one of the cells of 3×3 square grid is written a multiple of 101.
Note that 101 cannot be written in one of the cells, because in that case the product of eight positive integers needs to be 20 and their sum needs to be 114.
If the product of eight positive integers is 20, it means that at most 3 of them can be greater than 1 and at least the other five numbers are equal to 1.
On the other hand, each of them should be less than or equal to 20. Therefore, the sum of these eight numbers will be less than or equal to $20 + 20 + 20 + 1 + 1 + 1 + 1 + 1 = 65$. Hence, their sum cannot be equal to 114.
Taking the above mentioned into consideration, it follows that in one of the cells of 3×3 square grid is written a multiple of 101, less than 215.
Hence, we obtain that in one of the cells is written 202.
In this case, the sum of the other eight positive integers is 13 and their product is 10.
Therefore, these eight numbers are 1, 1, 1, 1, 1, 1, 2, 5.
Thus, there are six 1 written in the cells of 3×3 square grid. □

Problem 4.259. *Given that an interior angle of a regular n–gon is equal to $k°$, where k is a positive integer. What is the total number of all possible values of n?*

(A) 12 (B) 15 (C) 18 (D) 20 (E) 22

Solution. Answer. (E)
Recall that an interior angle of a regular n–gon is equal to
$$180° - \frac{360°}{n}.$$
According to the condition of the problem, we have that $n \geq 3$ and $n \mid 360$. Note that
$$360 = 2^3 \cdot 3^2 \cdot 5.$$
Therefore, the number of divisors of 360 is
$$(3+1)(2+1)(1+1) = 24.$$
As $n \geq 3$, so $n \neq 1$ and $n \neq 2$, hence the number of all possible values of n is $24 - 2 = 22$. □

Problem 4.260. *From ten consecutive positive integers were chosen seven numbers, such that their sum is equal to 2020. What is the greatest possible value of the sum of three not chosen numbers?*

(A) 600 (B) 700 (C) 875 (D) 880 (E) 890

Solution. Answer. (C)
Let these 10 consecutive positive integers be
$$n+1, n+2, ..., n+10,$$
where n is a nonnegative integer.
Assume that from these 10 numbers we did not choose the following 3 numbers
$$n+m, n+k, n+l,$$
where $m, k, l \in \{1, 2, ..., 10\}$.
According to the condition of the problem, we have that
$$10n + 55 - (3n + m + k + l) = 2020.$$
Thus, it follows that
$$7n = 1965 + (m+k+l).$$
Hence, we obtain that
$$3n + m + k + l = 3 \cdot \frac{1965 + m + k + l}{7} + m + k + l =$$
$$= \frac{3 \cdot 1965 + 10(m+k+l)}{7}.$$
Therefore, the sum of not chosen numbers is the greatest is $m+k+l$ is the greatest.
Note that $m+k+l$ leaves a remainder of 2 after division by 7. Recall that m, k, l are pairwise different and less than or equal to 10. Thus, the greatest possible value of $m+k+l$ is 23.
In this case, we obtain that
$$\frac{3 \cdot 1965 + 10(m+k+l)}{7} = 875.$$
□

Problem 4.261. *A rectangular prism is called "beautiful", if its three dimensions are positive integers. Let a rectangular prism M be divided by three planes parallel to its faces into eight "beautiful" rectangular prisms. Given that the volumes of four of these "beautiful" rectangular prisms are equal to 1, 2, 3, 5. What is the value of the total surface area of rectangular prism M?*

(A) 108   (B) 84   (C) 72   (D) 60   (E) 30

Solution. Answer. (A)
From the assumptions of the problem, it follows that the dimensions of four of these eight "beautiful" rectangular prisms are equal to

$$1 \times 1 \times 1,\ 1 \times 1 \times 2,\ 1 \times 1 \times 3,\ 1 \times 1 \times 5.$$

Thus, it follows that rectangular prism M has dimensions $3 \times 4 \times 6$.
Therefore, its total surface area is equal to

$$2(3 \cdot 4 + 4 \cdot 6 + 3 \cdot 6) = 108.$$

□

Problem 4.262. *Let 2 two-digit numbers be chosen at random. What is the probability of the event that the positive difference of these 2 two-digit numbers is also a two-digit number?*

(A) $\dfrac{36}{89}$ (B) $\dfrac{2}{5}$ (C) $\dfrac{72}{89}$ (D) $\dfrac{4}{5}$ (E) $\dfrac{9}{10}$

Solution. Answer. (C)
Note that the number of choice for choosing 2 two-digit numbers at random (from the list of all two-digit numbers) is equal to

$$\binom{90}{2} = 45 \cdot 89.$$

Assume that the positive difference of chosen 2 two-digit numbers is also a two-digit number.
Let us denote by x the smaller number among the chosen 2 numbers and denote by $n(x)$ the number of choices for choosing the greater number.
We have that
$n(x) = 0$, if $x \in \{90, 91, ..., 99\}$,
$n(x) = 90 - x$, if $x \in \{10, 11, ..., 89\}$.
Therefore, the number of favorable outcomes is equal to

$$1 + 2 + ... + 80 = 40 \cdot 81.$$

Hence, the probability of the event that the positive difference of chosen 2 two-digit numbers is also a two-digit number is equal to
$$\frac{40 \cdot 81}{45 \cdot 89} = \frac{72}{89}.$$

□

Problem 4.263. Let the graph of function $y = g(x)$ be symmetric to the graph of function $y = x^2$ with respect to point $M = (3, 9)$. What is the value of the length of the line segment, such that its endpoints are the intersection points of the graph of function $y = g(x)$ with the x−axis?

(A) 2 (B) 6 (C) $6\sqrt{2}$ (D) $6\sqrt{3}$ (E) 18

Solution. Answer. (C)
We have that $y = g(x)$ is a quadratic function and the vertex of its graph is the symmetric point of point $(0, 0)$ with repsect to point M, that is point $(6, 18)$. Thus, it follows that
$$g(x) = k(x - 6)^2 + 18.$$
Point M belongs to the graph of function $y = g(x)$, hence
$$9 = k \cdot 9 + 18.$$
We obtain that $k = -1$. Therefore, the intersection points of the graph of function $y = g(x)$ with the x−axis are the solutions of the following system of equation
$$\begin{cases} -(x-6)^2 + 18 = 0, \\ y = 0. \end{cases}$$
Note that the solutions of this system of equations are $(6 - 3\sqrt{2}, 0)$ and $(6 + 3\sqrt{2}, 0)$.
Therefore, the length of the line segment, such that its endpoints are the intersection points of the graph of function $y = g(x)$ with the x−axis is equal to
$$6 + 3\sqrt{2} - (6 - 3\sqrt{2}) = 6\sqrt{2}.$$
□

Problem 4.264. What is the total number of all three-digit numbers, not containing any zero digit, for which there is a digit such that after erasing that digit the obtained two-digit number is divisible by 3? For example three-digit numbers 121 and 123 satisfy these conditions.

(A) 120 (B) 159 (C) 729 (D) 540 (E) 513

Solution. Answer. (E)
A three-digit number that does not have any zero digit and does not satisfy the assumptions of the problem, we call an "insignificant" number.
Note that the total number of "insignificant" numbers, such that the sum of the digits for each of them is divisible by 3 is equal to
$$2 \cdot 3 \cdot 3 \cdot 3 = 54.$$
The total number of "insignificant" numbers, such that the sum of all digits for each of them leaves a remainder of 1 after division by 3 is equal to
$$3 \cdot 3 \cdot 3 \cdot 3 = 81.$$
The total number of "insignificant" numbers, such that the sum of all digits for each of them leaves a remainder of 2 after division by 3 is equal to
$$3 \cdot 3 \cdot 3 \cdot 3 = 81.$$
Therefore, the answer is
$$9 \cdot 9 \cdot 9 - 54 - 2 \cdot 81 = 513.$$
□

Problem 4.265. *What is the value of the product of all solutions of the following equation?*
$$3^{\sqrt{\log_3 x}} = 4^{\sqrt{\log_2^3 x}}.$$

(A) 1   (B) $2^{\sqrt{\log_2 \sqrt[4]{3}}}$ (C) 2   (D) 3   (E) 4

Solution. Answer. (B)
We have that
$$\log_2 3^{\sqrt{\log_3 x}} = \log_2 4^{\sqrt{\log_2^3 x}}.$$

We deduce that
$$\sqrt{\log_3 x} \cdot \log_2 3 = 2\sqrt{\log_2^3 x}.$$

Thus, it follows that
$$\begin{cases} x \geq 1, \\ \log_3 x \cdot \log_2 3 \cdot \log_2 3 = 4 \cdot \log_2^3 x. \end{cases}$$

We obtain that
$$\log_2 x (\log_2 3 - 4\log_2^2 x) = 0.$$

Therefore, either $x = 1$ or
$$x = 2^{-\sqrt{\log_2 \sqrt[4]{3}}},$$
or
$$x = 2^{\sqrt{\log_2 \sqrt[4]{3}}}.$$

As $x \geq 1$, we deduce that the solutions of the given equation are $x = 1$ and
$$x = 2^{\sqrt{\log_2 \sqrt[4]{3}}}.$$

Thus, it follows that
$$1 \cdot 2^{\sqrt{\log_2 \sqrt[4]{3}}} = 2^{\sqrt{\log_2 \sqrt[4]{3}}}.$$

□

Problem 4.266. *Let $ABCD$ be a rhombus, such that $\angle A = 45°$ and $AC = 13$. Assume that ray BD intersects the circumcircle of triangle ABC at point E. What is the value of the length of line segment DE?*

(A) 12   (B) 13   (C) $13\sqrt{2}$ (D) 20   (E) 26

Solution. Answer. (B)
Note that line BD is the perpendicular bisector of line segment AC, hence lige segment BE is the diameter of circumcircle of triangle ABC (see the figure).

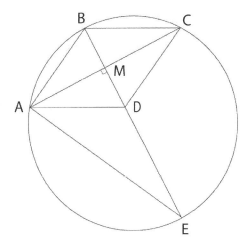

Thus, it follows that $\angle BAE = 90°$.

From triangle ABC according to law of sines, we obtain that
$$\frac{AC}{\sin 135°} = 2R = BE.$$

Hence, we deduce that $AC = \sqrt{2}R$.

Let $x = BM$. From triangle ABE we obtain that
$$\left(\frac{\sqrt{2}R}{2}\right)^2 = x(2R - x).$$

Thus, it follows that
$$x = R - \frac{\sqrt{2}R}{2},$$

and
$$DE = 2R - 2x = \sqrt{2}R = AC = 13.$$

$\square$

Problem 4.267. *Given that*
$$2009 = m + n + k,$$
where m, n, k are positive integers. At most with how many zeros can the product mnk end with?

(A) 5 (B) 6 (C) 7 (D) 8 (E) 10

Solution. Answer. (C)

Note that
$$2009 = 625 + 1000 + 384,$$
and the corresponding product
$$625 \cdot 1000 \cdot 384 = 240000000,$$
ends with 7 zeros.

Let us prove that the product mnk cannot end with eight or more zeros.
We proceed the proof by contradiction arguement. Assume that
$$10^8 \mid m \cdot n \cdot k.$$

Thus, it follows that
$$5^8 \mid mnk.$$
Therefore, two numbers among the numbers m, n, k are divisible by 625.
Without loss of generality one can assume that $625 \mid m$ and $625 \mid n$.
We have that
$$2^8 \mid mnk,$$
and $m \leq 2 \cdot 625, n \leq 2 \cdot 625$. Therefore $64 \mid k$.
$m + n$ is odd, thus $m + n = 3 \cdot 625$. We deduce that
$$k = 2009 - 3 \cdot 625 = 134.$$
Note that 134 is not divisible by 64.
Hence, the answer is 7. $\square$

Problem 4.268. *Let*
$$p(x) = x^4 + x^3 - 7x^2 - x + 6,$$
and $q(x)$ is a polynomial with real coefficients, such that $p(x) + q(x) \geq 0$ and $p(x)q(x) \leq 0$ for any value of x. What is the value of the expression $|q(-2) + q(4)|$.

(A) 0   (B) 200   (C) 198   (D) 10   (E) 12

Solution. Answer. (C)
Note that
$$p(x) = x^3(x+1) - (x+1)(7x-6) = (x+1)(x^3 - 7x + 6) =$$
$$= (x+1)(x(x^2-1) - 6(x-1)) = (x+1)(x-1)(x-2)(x+3).$$
From the condition $p(x)q(x) \leq 0$, it follows that $q(x) = p(x)r(x)$, where $r(x)$ is a polynomial with real coefficients and for any value of x we have that $r(x) \leq 0$.
We have that
$$p(x) + q(x) = p(x)(1 + r(x)) \geq 0.$$
We deduce that
$$1 + r(x) = p(x)s(x),$$
where $s(x) \geq 0$ for any value of x.
Hence, we obtain that
$$r(x) = p(x)s(x) - 1.$$
Therefore
$$s(x)p(x) \leq 1.$$
Let us prove that $s(x)$ is a zero polynomial. We proceed the proof by contradiction argument. Assume that $s(x)$ is a non zero polynomial, then for a large enough x_0 we have that $s(x_0) = a > 0$ and $p(x_0) > \dfrac{1}{a}$.
Thus, it follows that $s(x_0)p(x_0) > 1$. This leads to a contradiction. Therefore $s(x)$ is a zero polynomial.

We deduce that $q(x) = -p(x)$. Therefore
$$q(-2) + q(4) = -(p(-2) + p(4)) = -(-12 + 210) = -198,$$
and
$$|q(-2) + q(4)| = 198.$$
$\square$

Problem 4.269. *A committee of eight scientists consists of local scientists and visitor scientists. Every scientist brought two identical copies of a self-authored manuscript. Each of them exchanged the first copy of a self-authored manuscript with one copy of a self-authored manuscript of other attendee scientist and the second copy of a self-authored manuscript with one copy of self-authored manuscript of another attendee scientist. Given that any two scientists can exchange a copy of their self-authored manuscripts if one of them is a local scientist and another one is a visitor scientist. In how many different ways can they perform such exchanges?*

(A) 72   (B) 75   (C) 84   (D) 88   (E) 90

Solution. Answer. (E)
Let us denote the number of local scientists by m, then the total number of their self-authored manuscripts is $2m$ and the total number of self-authored manuscripts of visitor scientists is equal to $2(8-m)$. Given that
$$2m = 2(8-m).$$
Thus, it follows that $m = 4$.
Let us enumerate the local scientists by numbers 1, 3, 5, 7 and the visitor scientists by numbers 2, 4, 6, 8.
If the scientists with numbers i and j exchanged their self-authored manuscripts, then we denote that fact in following way
$$i \longleftrightarrow j.$$
We say that scientists with numbers $i_1, i_2, ..., i_k$, where $k \geq 3$, created a cycle with length k, if
$$i_1 \longleftrightarrow i_2, i_2 \longleftrightarrow i_3, ..., i_{k-1} \longleftrightarrow i_k, i_k \longleftrightarrow i_1.$$

According to the condition of the problem scientists need to create either a cycle of a length 8 or two cycles of lengths 4. In each cycle any two adjacent numbers have different parity.
Note that the number of the cycles of a length 8 is equal to
$$\frac{3!}{2!} \cdot 4! = 72.$$
The number of two cycles of lengths 4 is equal to $3 \cdot 6 = 18$.
Therefore, the total number of different ways of performing such exchanges is $72 + 18 = 90$. □

Problem 4.270. *Let u and v be positive numbers, such that $|u-v| \geq 1$. What is the smallest possible value of the expression $uv + \frac{u}{v} + \frac{v}{u}$?*

(A) 3   (B) 3.5   (C) 4   (D) $\sqrt{35}$ (E) 6

Solution. Answer. (C)
Note that
$$uv + \frac{u}{v} + \frac{v}{u} = uv + \frac{(u-v)^2}{uv} + 2 \geq uv + \frac{1}{uv} + 2 =$$
$$= \left(\sqrt{uv} - \frac{1}{\sqrt{uv}}\right)^2 + 4 \geq 4.$$
Thus, it follows that
$$uv + \frac{u}{v} + \frac{v}{u} \geq 4.$$

Note that, if
$$u = \frac{\sqrt{5}+1}{2},$$
and
$$v = \frac{\sqrt{5}-1}{2},$$
then $u - v = 1$ and $uv = 1$. Therefore
$$uv + \frac{u}{v} + \frac{v}{u} = uv + \frac{(u-v)^2}{uv} + 2 = 4.$$
Thus, the smallest possible value of the given expression is 4. □

Problem 4.271. *Consider a coordinate system O_{xyz}. Let solid φ consists of all points $M(x, y, z)$, such that for each of them the following inequality holds true:*
$$x^2 + y^2 + z^2 \leq |x| + |y| + |z|.$$
What is the value of the volume of solid φ?

(A) $\dfrac{\sqrt{3}\pi}{2}$ (B) $2\sqrt{3}\pi + 4$ (C) $\sqrt{3}\pi + 4$ (D) $\dfrac{4}{3}\pi$ (E) 1

Solution. Answer. (B)
Note that solid φ is symmetric with respect to each of planes O_{xy}, O_{yz}, O_{xz}. Let solid σ consists of all points $M(x, y, z)$, such that for each of them the following conditions holds true:
$$\begin{cases} x \geq 0, \\ y \geq 0, \\ z \geq 0, \\ \left(x - \dfrac{1}{2}\right)^2 + \left(y - \dfrac{1}{2}\right)^2 + \left(z - \dfrac{1}{2}\right)^2 \leq \left(\dfrac{\sqrt{3}}{2}\right)^2. \end{cases}$$

Taking into consideration the above mentioned, we obtain that the volume of solid φ is 8 times greater than the volume of solid σ. Note that solid σ is bounded by coordinate planes O_{xy}, O_{yz}, O_{xz} and by a part of a circumscribed sphere of the cube with the following vertices (with nonnegative coordinates):
$$(0,0,0), (0,0,1), (0,1,0), (1,0,0), (1,1,0), (1,0,1), (0,1,1), (1,1,1).$$
The volume of solid σ is:
$$\frac{1}{2} \cdot \left(\frac{4\pi}{3} \cdot \left(\frac{\sqrt{3}}{2}\right)^3 - 1\right) + 1 = \frac{\sqrt{3}\pi}{4} + \frac{1}{2}.$$
Therefore, the volume of solid φ is equal to $2\sqrt{3}\pi + 4$. □

Problem 4.272. *Given that*
$$\sum_{n=1}^{\infty} \frac{\cos(n\phi)}{2^n} = \sum_{n=1}^{\infty} \frac{\sin(n\phi)}{2^n}.$$

What is the value of $\sin(2\phi)$?

(A) 1 (B) $\frac{3}{4}$ (C) $\frac{1}{2}$ (D) $\frac{1}{4}$ (E) 0

Solution. Answer. (B)

We have that
$$\cos\phi \cdot \sum_{n=1}^{\infty} \frac{\cos(n\phi)}{2^n} = \cos\phi \cdot \sum_{n=1}^{\infty} \frac{\sin(n\phi)}{2^n},$$

$$\sum_{n=1}^{\infty} \frac{\cos(n\phi) \cdot \cos\phi}{2^n} = \sum_{n=1}^{\infty} \frac{\sin(n\phi) \cdot \cos\phi}{2^n},$$

$$\sum_{n=1}^{\infty} \frac{\cos((n+1)\phi) + \cos((n-1)\phi)}{2^{n+1}} = \sum_{n=1}^{\infty} \frac{\sin((n+1)\phi) + \sin((n-1)\phi)}{2^{n+1}}.$$

$$\sum_{n=1}^{\infty} \frac{\cos((n+1)\phi)}{2^{n+1}} + \frac{1}{2^2}\sum_{n=1}^{\infty} \frac{\cos((n-1)\phi)}{2^{n-1}} =$$

$$= \sum_{n=1}^{\infty} \frac{\sin((n+1)\phi)}{2^{n+1}} + \frac{1}{2^2}\sum_{n=1}^{\infty} \frac{\sin((n-1)\phi)}{2^{n-1}},$$

$$\sum_{n=1}^{\infty} \frac{\cos(n\phi)}{2^n} + \frac{1}{2^2}\sum_{n=1}^{\infty} \frac{\cos(n\phi)}{2^n} + \frac{1}{4} - \frac{\cos\phi}{2} =$$

$$= \sum_{n=1}^{\infty} \frac{\sin(n\phi)}{2^n} + \frac{1}{4}\sum_{n=1}^{\infty} \frac{\sin(n\phi)}{2^n} - \frac{\sin\phi}{2},$$

$$\frac{1}{4} - \frac{\cos\phi}{2} = -\frac{\sin\phi}{2}.$$

Thus, it follows that
$$\cos\phi - \sin\phi = \frac{1}{2}.$$

Hence, we obtain that
$$(\cos\phi - \sin\phi)^2 = \frac{1}{4}.$$

We deduce that
$$\sin 2\phi = 1 - \frac{1}{4} = \frac{3}{4}.$$

□

Problem 4.273. *Given point M on side AB and point N on side BC of triangle ABC, such that*
$$\frac{BM}{CN} = \frac{BN}{AM} = \frac{3}{5},$$
and
$$\angle BMN - \angle BNM = 60°.$$
What is the value of $\frac{MN}{AC}$?

(A) $\frac{3}{5}$ (B) $\frac{2}{5}$ (C) $\frac{3}{7}$ (D) $\frac{5}{7}$ (E) $\frac{1}{2}$

Solution. Answer. (C)
Let us consider parallelogram $MNCK$.
We denote
$$BM = 3x, BN = 3y, MN = 3z,$$
then
$$CN = 5x, AM = 5y.$$

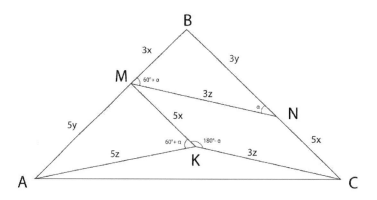

Note that
$$\angle AMK = \angle NBM,$$
and
$$\frac{AM}{BN} = \frac{MK}{BM}.$$
Hence, we deduce that
$$\triangle AMK \sim \triangle NBM.$$
Thus, it follows that
$$\frac{MN}{AK} = \frac{3}{5}.$$
Therefore
$$\angle AKC = 360° - (60° + \alpha + 180° - \alpha) = 120°.$$
From triangle AKC according to law of cosines, we obtain that
$$AC^2 = 9z^2 + 25z^2 + 3z \cdot 5z = 49z^2.$$
We deduce that
$$\frac{MN}{AC} = \frac{3z}{7z} = \frac{3}{7}.$$

Problem 4.274. Let m and n be positive integers, such that $n \mid (m + 20)$ and $m \mid (n + 21)$. What is the value of the sum of all digits of the greatest possible value of $m + n$?

  (A) 3   (B) 4   (C) 5   (D) 10   (E) 12

Solution. Answer. (B)
Let us consider the following two cases.
Case 1. If $m \leq n$.
If $n \leq 20$, then
$$m + n \leq 2n \leq 40.$$
Thus, it follows that $m + n \leq 40$.
If $n > 20$, then
$$m + 20 < m + n \leq 2n.$$
From the following conditions
$$m + 20 < 2n,$$
and
$$n \mid (m + 20),$$
we obtain that
$$m + 20 = n.$$
Hence, from the condition
$$m \mid (n + 21),$$
we obtain that $m \mid (m + 41)$.
Therefore
$$m \mid 41,$$
$$n = m + 20 \leq 61.$$
Thus, it follows that
$$m + n \leq 102.$$
Case 2. If $m > n$.
If $m \leq 21$, then
$$m + n < 2m \leq 42.$$
If $m > 21$, then
$$n + 21 < m + n < 2m.$$
From the following conditions
$$n + 21 < 2m,$$
and
$$m \mid (n + 21),$$
we obtain that
$$m = n + 21.$$
We have that
$$n \mid (m + 20).$$
Therefore
$$n \mid (n + 41).$$
Hence $n \mid 41$. Thus, it follows that
$$\begin{cases} n \leq 41, \\ m = n + 21 \leq 62. \end{cases}$$

We obtain that
$$m + n \leq 103.$$

Therefore, for both cases we obtain that the greatest possible value of $m + n$ is 103. Note that when $m = 62$ and $n = 41$, we have that $m + n = 103$.
Hence, the answer is $1 + 0 + 3 = 4$. □

Problem 4.275. *Let the inradius of triangle ABC be equal to 1. Given that $\angle BAC = 30°$ and $\angle ABC = 45°$. A random segment of length 1 is drawn inside of triangle ABC, such that its projections on sides BC, AC, AB are a, b, c, respectively. What is the probability of the event that*
$$\sqrt{2}a + 2b = (\sqrt{3} + 1)c?$$

(A) $\dfrac{1}{3}$ (B) $\dfrac{5}{12}$ (C) $\dfrac{2}{3}$ (D) $\dfrac{7}{12}$ (E) $\dfrac{3}{4}$

Solution. Answer. (D)
Let MN be the diameter of the incircle of triangle ABC, such that MN is perpendicular to AB (see the figure). Without loss of generality one can assume that considered random segment of length 1 is line segment IX, where I is the incenter of triangle ABC and X is a point on the left semicircle.

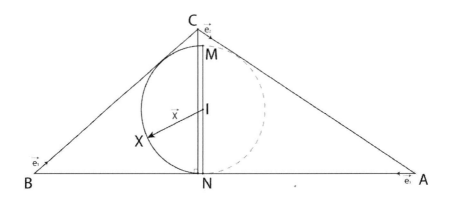

Let h be the length of the altitude drawn from vertex C. Then
$$BC = h\sqrt{2}, \quad AC = 2h, \quad AB = (\sqrt{3} + 1)h.$$

We have that
$$\overrightarrow{AB} + \overrightarrow{BC} + \overrightarrow{CA} = \vec{0}.$$

Thus, it follows that
$$\sqrt{2}\vec{e}_1 + 2\vec{e}_2 + (\sqrt{3} + 1)\vec{e}_3 = \vec{0},$$

where
$$|\vec{e}_1| = |\vec{e}_2| = |\vec{e}_3| = 1.$$

Hence, we obtain that
$$\sqrt{2}\vec{e}_1 \cdot \vec{x} + 2\vec{e}_2 \cdot \vec{x} + (\sqrt{3} + 1)\vec{e}_3 \cdot \vec{x} = 0,$$
$$\pm\sqrt{2}a \pm 2b + (\sqrt{3} + 1)c = 0.$$

We deduce that
$$\sqrt{2}a + 2b = (\sqrt{3} + 1)c,$$

if and only if the vector $\vec{x}$ is located in the following angle (see the figure).

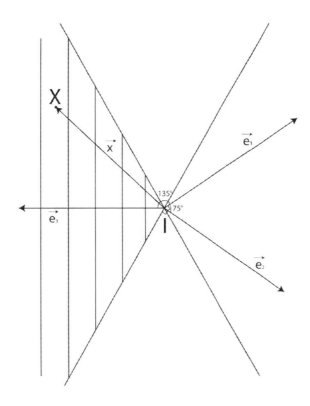

Therefore, the probability of the event that $\sqrt{2}a + 2b = (\sqrt{3}+1)c$ is:

$$\frac{105}{180} = \frac{7}{12}.$$

□

4.12 Solutions of AMC 12 type practice test 12

Problem 4.276. *What is the value of the sum of the opposite of 2020 and the reciprocal of $\frac{1}{2019}$?*

(A) $-\frac{1}{2019 \cdot 2020}$ (B) $2019\frac{1}{2020}$ (C) $-2020\frac{1}{2019}$ (D) -1 (E) 1

Solution. Answer. (D)

Note that the opposite of 2020 is -2020 and the reciprocal of $\frac{1}{2019}$ is 2019.

Thus, it follows that the the sum of the opposite of 2020 and the reciprocal of $\frac{1}{2019}$ is:

$$-2020 + 2019 = -1.$$

□

Problem 4.277. *Boat tour ticket costs 25$ for an adult and 10$ for a child. Given that a group of 7 people paid 130$ for 7 tickets. How many children are in this group?*

(A) 1 (B) 3 (C) 4 (D) 6 (E) 7

Solution. Answer. (B)

Let x be the number of children in this group. Therefore, the number of adults is $7 - x$. Given that

$$10x + 25(7 - x) = 130.$$

Thus, it follows that
$$175 - 15x = 130.$$

We deduce that $x = 3$.

□

Problem 4.278. *Let one of the numbers 1,2,...,9 be equal to the arithmetic mean of the other eight numbers. What is the value of the sum of all digits of the sum of the other eight numbers?*

(A) 2 (B) 3 (C) 4 (D) 6 (E) 11

Solution. Answer. (C)

Let $n \in \{1, 2, ..., 9\}$ and n is equal to the arithmetic mean of the other eight numbers.
Thus, it follows that
$$n = \frac{1 + 2 + ... + 9 - n}{8}.$$

We deduce that $9n = 45$. We obtain that $n = 5$.
Therefore, the sum of these eight numbers is

$$1 + 2 + ... + 9 - 5 = \frac{9 \cdot 10}{2} - 5 = 40.$$

Hence, the sum of all digits is $4 + 0 = 4$.

□

Problem 4.279. *Let n be any positive integer. What is the number of elements of the set*

$$\left\{1, \frac{1}{2}, \frac{1}{3}, ..., \frac{1}{n}, ...\right\},$$

such that each of them is a solution of the following inequality?

$$100x^2 - 25x + 1 < 0.$$

(A) 13   (B) 14   (C) 15   (D) 16   (E) 19

Solution. Answer. (B)
Note that the roots of the quadratic trinomial $100x^2 - 25x + 1$ are $\frac{1}{20}$ and $\frac{1}{5}$. Thus, it follows that the set of the solutions of the inequality

$$100x^2 - 25x + 1 < 0,$$

is the interval

$$\left(\frac{1}{20}, \frac{1}{5}\right).$$

We need to find the total number of all positive integers n, such that

$$\frac{1}{20} < \frac{1}{n} < \frac{1}{5}.$$

Thus, it follows that
$$5 < n < 20.$$

Hence, we obtain that
$$n \in \{6, 7, ..., 19\}.$$

Therefore, the answer is $19 - 5 = 14$. □

Problem 4.280. *The first shop sells 450 grams of some type of candy for 5\$ and the second shop sells 500 grams of the same type of candy for 6\$. How much more in percentage is the price of the candy in the second shop compared to the first shop?*

(A) 12   (B) 10   (C) 9   (D) 8   (E) 5

Solution. Answer. (D)
Given that in the first shop 50 grams of the candy costs $\frac{5}{9}$\$ and in the second shop 50 grams of the same candy costs $\frac{3}{5}$\$.
Note that in order to find how much more in percentage is the candy in the second shop compared to the first shop, we need to calculate the value of the following expression

$$\frac{\frac{3}{5} - \frac{5}{9}}{\frac{5}{9}} \cdot 100\% = 8\%.$$

□

Problem 4.281. *For which value of m do the graphs of functions $y = x - 1$, $y = 3x - 5$, $y = mx - 41$ intersect at one point?*

(A) 5   (B) 10   (C) 19   (D) 21   (E) 22

Solution. Answer. (D)
At first, let us find the intersection point $M(x_0, y_0)$ of the graphs of functions $y = x - 1$ and $y = 3x - 5$. We have that
$$\begin{cases} y_0 = x_0 - 1, \\ y_0 = 3x_0 - 5. \end{cases}$$
Thus, it follows that
$$x_0 = 2, y_0 = 1.$$
According to the condition of the problem the line given by the equation $y = mx - 41$ passes through point $M(2, 1)$. Hence, we obtain that
$$2m - 41 = 1.$$
Therefore $m = 21$. □

Problem 4.282. *What is the value of the sum of all real solutions of the following equation?*
$$6^x - 2 \cdot 3^x - 27 \cdot 2^x + 54 = 0.$$

(A) 2   (B) 3   (C) 3.5   (D) 4   (E) 6

Solution. Answer. (D)
Let us rewrite given equation in the following way
$$3^x(2^x - 2) - 27(2^x - 2) = 0.$$
Thus, it follows that
$$(2^x - 2)(3^x - 27) = 0.$$
We deduce that, either $2^x - 2 = 0$ or $3^x - 27 = 0$.
Therefore, all possible real solutions of given equation are $x = 1$ and $x = 3$ and their sum is equal to 4. □

Problem 4.283. *What is the greatest possible solution of the following equation?*
$$\log_2 x = \sqrt{\log_3 x}.$$

(A) 1   (B) $2^{\log_3 2}$ (C) 2   (D) 3   (E) 3.2

Solution. Answer. (B)
We have that
$$\log_3 x = \frac{\log_2 x}{\log_2 3}.$$
Thus, it follows that
$$\log_2 x = \frac{\sqrt{\log_2 x}}{\sqrt{\log_2 3}}.$$
We deduce that
$$\sqrt{\log_2 x}(\sqrt{\log_2 x} - \sqrt{\log_3 2}) = 0.$$

Therefore, either
$$\sqrt{\log_2 x} = 0,$$
or
$$\sqrt{\log_2 x} - \sqrt{\log_3 2} = 0.$$

We deduce that
$$\log_2 x = \log_3 2.$$

Thus, it follows that
$$x = 2^{\log_3 2}.$$

Therefore, the solutions of given equation are 1 and $2^{\log_3 2}$. Note that
$$\log_3 2 > 0.$$

Hence, we obtain that
$$2^{\log_3 2} > 2^0 = 1.$$

Thus, the greatest solution of given equation is $2^{\log_3 2}$. $\square$

Problem 4.284. *Given three pairwise different positive integers, such that the sum of each two of them is greater than the consecutive number of the third number. What is the smallest possible value of the sum of these three numbers?*

(A) 10   (B) 11   (C) 12   (D) 13   (E) 14

Solution. Answer. (C)
Let a, b, c be positive integers, such that $a < b < c$ and the sum of each two of them is greater than the consecutive number of the third number.
Thus, it follows that
$$a + b > c + 1.$$

We have that $c - b \geq 1$, therefore
$$a > c - b + 1 \geq 2.$$

We deduce that $a \geq 3$.
Thus, it follows that
$$a \geq 3, b \geq 4, c \geq 5.$$

We obtain that
$$a + b + c \geq 12.$$

Note that 3, 4, 5 satisfy the assumptions of the problem, therefore the answer is $3 + 4 + 5 = 12$. $\square$

Problem 4.285. *What is the value of the following expression?*

$$\frac{3}{1 \cdot 2} - \frac{5}{2 \cdot 3} + \frac{7}{3 \cdot 4} - \frac{9}{4 \cdot 5} + \frac{11}{5 \cdot 6} - \frac{13}{6 \cdot 7} + \frac{15}{7 \cdot 8} - \frac{17}{8 \cdot 9} + \frac{19}{9 \cdot 10}.$$

(A) 1.1   (B) 0.9   (C) 0.75   (D) 0.6   (E) 0.5

Solution. Answer. (A)
Note that
$$\frac{2k+1}{k(k+1)} = \frac{(k+1)+k}{k(k+1)} =$$
$$\frac{k+1}{k(k+1)} + \frac{k}{k(k+1)} = \frac{1}{k} + \frac{1}{k+1}.$$

Thus, it follows that
$$\frac{2k+1}{k(k+1)} = \frac{1}{k} + \frac{1}{k+1},$$
where $k = 1, 2, \ldots$ Using the last equation let us rewrite the given expression in the following way

$$\frac{1}{1} + \frac{1}{2} - \left(\frac{1}{2} + \frac{1}{3}\right) + \frac{1}{3} + \frac{1}{4} - \left(\frac{1}{4} + \frac{1}{5}\right) + \frac{1}{5} + \frac{1}{6} - \left(\frac{1}{6} + \frac{1}{7}\right) +$$
$$+ \frac{1}{7} + \frac{1}{8} - \left(\frac{1}{8} + \frac{1}{9}\right) + \frac{1}{9} + \frac{1}{10} = 1 + \frac{1}{10} = 1.1.$$

$\square$

Problem 4.286. *Let a, b, c, d be such digits that $\overline{ab}, \overline{cd}, \overline{ac}, \overline{bd}$ are two-digit numbers. Given that*
$$\overline{ab} + \overline{cd} = \overline{ac} + \overline{bd}.$$
What is the total number of all possible quadruples (a, b, c, d)?

(A) 810   (B) 729   (C) 700   (D) 500   (E) 100

Solution. Answer. (A)
We have that
$$10a + b + 10c + d = 10a + c + 10b + d.$$
Thus, it follows that $c = b$.
Therefore, we need to find the number of quadruples of the form (a, b, b, d), where $a \in \{1, 2, \ldots, 9\}$, $b \in \{1, 2, \ldots, 9\}$ and $d \in \{0, 1, 2, \ldots, 9\}$.
According to the product rule of counting the number of all such quadruples is $9 \cdot 9 \cdot 10 = 810$. $\square$

Problem 4.287. *Let $A(a, b)$ and $B(a - b, a)$ be two points on the coordinate plane, where a and b are positive integers and $a > b$. Let O be the point $(0, 0)$ and the area of triangle ABO be 96. What is the value of $a + b$?*

(A) 28   (B) 27   (C) 26   (D) 25   (E) 24

Solution. Answer. (E)
Let us consider the figure below.

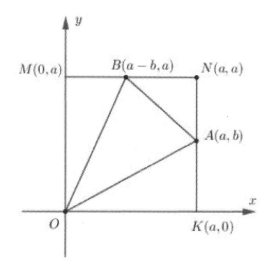

We have that
$$(OAB) = (OMNK) - (OMB) - (BNA) - (OAK) =$$
$$= a^2 - \frac{a(a-b)}{2} - \frac{b(a-b)}{2} - \frac{ab}{2} = \frac{a^2 - ab + b^2}{2}.$$
According to the condition of the problem, we have that
$$a^2 - ab + b^2 = 192.$$
Thus, it follows that
$$a^2 + b^2 + (a-b)^2 = 384 = 3 \cdot 2^7.$$
Taking into consideration that the square of any odd number leaves a remainder of 1 after division by 4, we deduce that $8 \mid a$ and $8 \mid b$. Taking this into consideration and using the last equation, one can easily obtain that
$$a = 16, b = 8.$$
Hence, we obtain that
$$a + b = 24.$$
□

Problem 4.288. *Let two cars start moving simultaneously toward each other from places A and B. Given that cars move with constant speeds and they cross each other in 2 hours. Given also that if these two cars start moving simulatenously toward each other from places A and B, such that each car moves with a constant speed that is greater by 10 km per hour than its initial speed, then they cross each other in 1 hour 48 minutes. What is the value of the distance (in kilometres) between A and B?*

(A) 300 (B) 360 (C) 400 (D) 410 (E) 420

Solution. Answer. (B)
Let the distance between A and B be x km.
Note that the sum of the initial speeds of the cars is equal to $\frac{x}{2}$ km per hour. On the other hand, the sum of the initial speeds of the cars is equal to $\left(\frac{x}{1.8} - 20\right)$ km per hour.
Thus, it follows that
$$\frac{x}{2} = \frac{x}{1.8} - 20.$$
Hence, we obtain that $x = 360$. □

Problem 4.289. *Let two circles of radii 6 be tangent to each other, such that each of these circles is tangent to three sides of trapezoid ABCD with bases AD and BC. Given that AB = 13 and CD = 20. What is the value of the area of trapezoid ABCD?*

(A) 198 (B) 396 (C) 300 (D) 342 (E) 210

Solution. Answer. (D)
Let us consider the figure below.

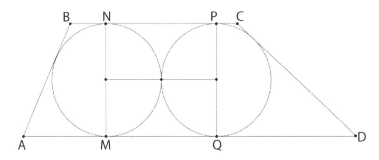

We have that
$$Area(ABCD) = Area(ABNM) + Area(MNPQ) + Area(QPCD) =$$
$$= Area(ABNM) + Area(QPCD) + 12 \cdot 12.$$

Note that if we connect trapezoids $ABNM$ and $PCDQ$, such that line segments MN and PQ coincide, then the area of the obtained trapezoid is equal to
$$(AB + CD) \cdot 6 = 198.$$

Thus, it follows that
$$Area(ABNM) + Area(QPCD) = 198.$$

Therefore, we obtain that
$$Area(ABCD) = 198 + 144 = 342.$$

$\square$

Problem 4.290. *Let ABC be a right triangle, such that $\angle B = 15°$ and the sum of the lenghts of legs AC, BC is equal to 10. Let CH be an altitude in triangle ABC and HE, HF be altitudes in triangles ACH, BCH, respectively. What is the value of the perimeter of rectangle $HECF$?*

(A) 10   (B) 8   (C) 7   (D) 6   (E) 5

Solution. Answer. (E)

Let M be the midpoint of leg BC (see the figure).

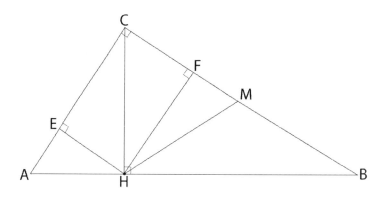

We have that $HM = MB$, therefore
$$\angle HMF = 2\angle HBC = 30°.$$

Thus, it follows that
$$HF = \frac{1}{2}HM = \frac{1}{4}BC.$$

In a similar way, we obtain that
$$HE = \frac{1}{4}AC.$$

Hence, the perimeter of rectangle $HECF$ is equal to
$$2(HE + HF) = \frac{1}{2}(AC + BC) = 5.$$

$\square$

Problem 4.291. *At most how many chess knights is possible to put on 4×4 grid square, such that each chess knight keeps under attack not more than two chess knights?*

(A) 10 (B) 11 (C) 12 (D) 13 (E) 14

Solution. Answer. (C)

At first, let us provide an example of an arrangment of 12 chess knights on 4×4 square grid, such that each chess knight keeps under attack not more than 2 chess knights (see the figure).

Now, let us prove that among 13 chess knights there is a knight, such that it keeps under attack more than 2 chess knights. Note that among 13 chess knights there are 4 chess knights that are in the cells with the same numbers (see the figure).

1	4	1	2
3	2	3	4
4	1	4	3
2	3	2	1

This ends the proof of the statement. Therefore, the answer is 12. □

Problem 4.292. *Given a square ABCD. Let M be a randomly chosen point in the interior part of square ABCD. What is the probability of the event that $\angle BMD \geq 135°$?*

(A) $\dfrac{\pi - 2}{2}$ (B) $\dfrac{1}{2}$ (C) $\dfrac{\pi}{8}$ (D) $\dfrac{1}{4}$ (E) $\dfrac{\pi}{15}$

Solution. Answer. (A)

Let φ be geometric location of all points M belonging to the interior part of square $ABCD$, such that
$$\angle BMD \geq 135°.$$

Let us take $AB = a$ and calculate the area of figure φ (see the figure below).

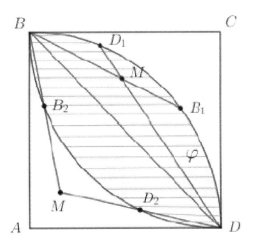

Consider circles with centers A and C, such that their radii are equal to a.
Let us prove that φ is the common part of these circles.
Note that, if M belongs to the common part of these circles, then

$$\angle BMD = \frac{270° + arc(B_1 D_1)}{2} \geq 135°.$$

If point M does not belong to the common part of these circles, then

$$\angle BMD = \frac{270° - arc(B_2 D_2)}{2} < 135°.$$

Therefore, the area of figure φ is equal to

$$2\left(\frac{\pi \cdot a^2}{4} - \frac{a \cdot a}{2}\right) = \frac{a^2}{2}(\pi - 2).$$

Thus, the answer is

$$\frac{\frac{a^2}{2}(\pi - 2)}{a^2} = \frac{\pi - 2}{2}.$$

□

Problem 4.293. *Let figure Φ consists of a cylinder and a cone with the same base (see the figure). Given that the radius of the base is 5, the altitude of the cylinder is 10 and the altitude of the cone is 12. What is the value of the total surface area of figure Φ?*

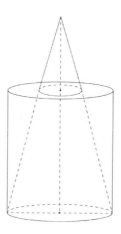

(A) 140π (B) 150π (C) $150\frac{1}{9}\pi$ (D) $151\frac{1}{9}\pi$ (E) 160π

Solution. Answer. (D)
At first, let us find the radius of the small cone that is outside of the cylinder (see the figure).

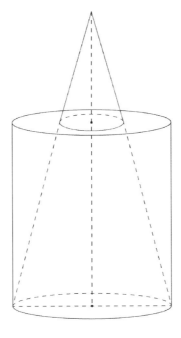

Note that the axial section of this small cone and the axial section of the initial cone create two similar triangles, therefore
$$\frac{12}{2} = \frac{5}{r}.$$
Thus, it follows that
$$r = \frac{5}{6}.$$
The slant height of the small cone is equal to
$$\frac{1}{6} \cdot \sqrt{12^2 + 5^2} = \frac{13}{6}.$$
Let S be the total surface area of figure Φ, then we obtain that
$$S = \pi \cdot \frac{5}{6} \cdot \frac{13}{6} + 25\pi - \frac{25}{36}\pi + 2\pi \cdot 5 \cdot 10 + 25\pi = 151\frac{1}{9}\pi.$$

$\square$

Problem 4.294. *Let the sum of four pairwise different integers be equal to 2022. Given that the positive difference of each two of these numbers is prime number. What is the greatest number among these four integers?*

(A) 409   (B) 507   (C) 508   (D) 509   (E) 510

Solution. Answer. (D)
Let m, n, p, k be given four integers, such that $m < n < p < k$ and
$$m + n + p + k = 2022,$$
and
$$n - m, p - m, k - m, p - n, k - n, k - p,$$
are prime numbers.
Note that at most two numbers among integers m, n, p, k can have the same parity.

Otherwise, the difference of the greatest and the smallest numbers among them is not less than 4 and is an even number. Thus, it follows that it cannot be prime number. This leads to a contradiction. Therefore, two among these numbers are even and two of them are odd.

As their positive difference is prime number, then the positive difference of the numbers with the same parity is equal to 2.

We obtain that these numbers have the following form

$$m, m+2, p, p+2.$$

Note that two among the numbers m, n, p, k leave the same remainder after a division by 3. Therefore, their difference is divisible by 3. On the other hand, their difference is prime number. Hence, their difference is equal to 3.

We deduce that these numbers have the following form

$$m, m+2, m+5, m+7.$$

According to the condition of the problem we have that

$$4m + 14 = 2022.$$

Thus, it follows that $m = 502$. The greatest number among these four numbers is $m + 7$ and $m + 7 = 509$. $\square$

Problem 4.295. *Let Φ be a figure on two dimensional coordinate plane consisting of points (x, y), such that*

$$|x^2 + y^2 - 1| + |(x-1)^2 + y^2 - 1| \leq 3.$$

What is the value of the area of figure Φ?

(A) π (B) 2π (C) 2.25π (D) 10   (E) $10 + \dfrac{\pi}{4}$

Solution. Answer. (C)

Note that given inequality can be rewritten in the following way.

$$\begin{cases} x^2 + y^2 - 1 \geq 0, \\ (x-1)^2 + y^2 - 1 \geq 0, \\ x^2 + y^2 - 1 + (x-1)^2 + y^2 - 1 \leq 3, \end{cases}$$

or

$$\begin{cases} x^2 + y^2 - 1 \geq 0, \\ (x-1)^2 + y^2 - 1 \leq 0, \\ x^2 + y^2 - 1 - (x-1)^2 - y^2 + 1 \leq 3, \end{cases}$$

or

$$\begin{cases} x^2 + y^2 - 1 \leq 0, \\ (x-1)^2 + y^2 - 1 \geq 0, \\ -x^2 - y^2 + 1 + (x-1)^2 + y^2 - 1 \leq 3, \end{cases}$$

or

$$\begin{cases} x^2 + y^2 - 1 \leq 0, \\ (x-1)^2 + y^2 - 1 \leq 0, \\ -x^2 - y^2 + 1 - (x-1)^2 - y^2 + 1 \leq 3. \end{cases}$$

The solutions of these system of inequalities are provided in the following figures.

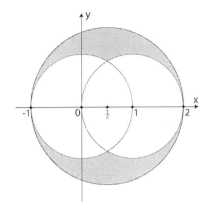
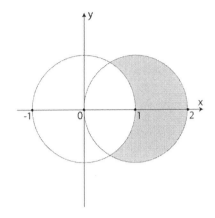
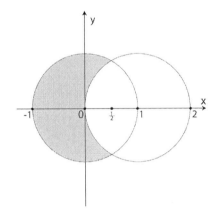
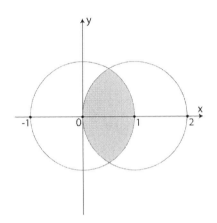

Therefore, the figure Φ is a circle with center $\left(\frac{1}{2}, 0\right)$ and radius $R = \frac{3}{2}$.
Thus, the answer is
$$\pi \cdot \frac{9}{4} = 2.25\pi.$$
$\square$

Problem 4.296. *Let in a 2013×2013 square grid, some of the vertices of some of the cells be marked in red. Given that each of the cells of the square grid has at least two red vertices. What is the smallest possible number of red vertices?*

(A) 1006^2 (B) $2 \cdot 1007^2$ (C) 2000^2 (D) 1000 (E) 2000

Solution. Answer. (B)
Let us note that in given 2013×2013 square grid we can choose $1007 \cdot 1007$ cells, such that any two of them have no common vertex (see the figure).

249

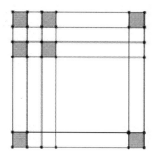

Each cell has at least two red vertices, therefore the number of red vertices is not less than $2 \cdot 1007^2$. Now, let us provide an example, such that the number of red vertices is $2 \cdot 1007^2$ (see the figure).

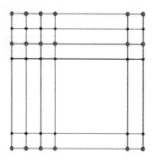

Therefore, the answer is $2 \cdot 1007^2$. □

Problem 4.297. *An ant moves from the bottom left corner to the top right corner of 6×8 rectangular grid. It can move only on the sides of unit cells, such that one move is either going up 1 unit or going down 1 unit or going right 1 unit. Given that the ant cannot pass twice the same side of any unit cell. What is the probability of the event that the ant passes through the center of symmetry of given 6×8 rectangular grid?*

(A) $\dfrac{18}{49}$ (B) $\dfrac{3}{7}$ (C) $\dfrac{31}{49}$ (D) $\dfrac{5}{7}$ (E) $\dfrac{6}{7}$

Solution. Answer. (C)

Note that each route of the ant in getting from the starting point to the destination is possible to represent as an eight-digit number $\overline{a_1 a_2 \ldots a_8}$, where $a_i \in \{1, 2, \ldots, 7\}, i = 1, 2, \ldots, 8$ and a_i is equal to the number of cells located under i^{th} horizontal step plus 1. For example, the eight-digit number corresponding to the figure below is 33422564.

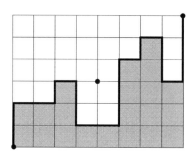

Note that the number of all possible routes is
$$7 \cdot 7 \cdot 7 \cdot 7 \cdot 7 \cdot 7 \cdot 7 \cdot 7 = 7^8.$$

Note that the route does not pass through the center of symmetry of given 6×8 rectangular grid if and only if $a_4, a_5 \in \{1, 2, 3\}$ or $a_4, a_5 \in \{5, 6, 7\}$.
Therefore, the probability of the event that the route of the ant does not pass through the center of symmetry of given 6×8 rectangular grid is
$$\frac{7^6(3 \cdot 3 + 3 \cdot 3)}{7^8} = \frac{18}{49}.$$

Thus, it follows that the probability of the event that the route of the ant passes through the center of symmetry of given 6×8 rectangular grid is
$$1 - \frac{18}{49} = \frac{31}{49}.$$

$\square$

Problem 4.298. *What is the value of the sum of the smallest and the greatest solutions of the following equation belonging to $\left[0, \frac{3\pi}{2}\right]$?*
$$\cos x - \sin x + \cos 4x = -0.5.$$

(A) $\dfrac{2\pi}{3}$ (B) π (C) $\dfrac{4\pi}{3}$ (D) $\dfrac{3\pi}{2}$ (E) $\dfrac{5\pi}{3}$

Solution. Answer. (D)

Note that if x_0 is a solution of given equation, then $\dfrac{3\pi}{2} - x_0$ is also a solution of given equation.
We have that
$$\cos\left(\frac{3\pi}{2} - x_0\right) - \sin\left(\frac{3\pi}{2} - x_0\right) + \cos 4\left(\frac{3\pi}{2} - x_0\right) =$$
$$= (-\sin x_0) - (-\cos x_0) + \cos(6\pi - 4x_0) =$$
$$= \cos x_0 - \sin x_0 + \cos 4x_0 = -0.5.$$

This ends the proof of the statement. Hence, if x_0 is a solution of given equation, then $\dfrac{3\pi}{2} - x_0$ is also a solution of given equation.
If x_1 is the smallest solution of given equation belonging to $\left[0, \frac{3\pi}{2}\right]$, then $x_1 \neq \dfrac{3\pi}{2} - x_1$ and $\dfrac{3\pi}{2} - x_1$ is the greatest solution of given equation belonging to $\left[0, \frac{3\pi}{2}\right]$.
Therefore, the sum of the smallest and the greatest solutions of given equation belonging to $\left[0, \frac{3\pi}{2}\right]$ is:
$$x_1 + \left(\frac{3\pi}{2} - x_1\right) = \frac{3\pi}{2}.$$

$\square$

Problem 4.299. *Let n be a positive integer and sequence x_n be defined as follows:*
$$x_1 = 1, x_2 = 2, x_3 = 5,$$
and
$$x_{n+3} = \frac{x_{n+2}^2 - x_{n+2} x_{n+1}}{x_{n+1} - x_n},$$
where $n = 1, 2, \ldots$. What is the total number of all possible values of n, such that for each of them $\frac{x_{n+1}}{x_n}$ is an integer?

(A) 1   (B) 2   (C) 4   (D) 2016   (E) 2017

Solution. Answer. (B)
Note that
$$\frac{x_{n+3}}{x_{n+2}} = \frac{x_{n+2} - x_{n+1}}{x_{n+1} - x_n}.$$
Thus, it follows that
$$\frac{x_{n+2}}{x_{n+1}} = \frac{x_{n+1} - x_n}{x_n - x_{n-1}}.$$
...
$$\frac{x_4}{x_3} = \frac{x_3 - x_2}{x_2 - x_1}.$$
Multiplying all these equations, we obtain that
$$\frac{x_{n+3}}{x_3} = \frac{x_{n+2} - x_{n+1}}{x_2 - x_1},$$
Thus, it follows that
$$x_{n+3} = 5(x_{n+2} - x_{n+1}),$$
where $n = 0, 1, 2, \ldots$. We have that
$$x_1 = 1, x_2 = 2, x_3 = 5, x_4 = 15, x_5 = 50, x_6 = 175, \ldots$$
Let us prove by mathematical induction that
$$3 < \frac{x_{n+1}}{x_n} < 4,$$
where $n = 4, 5, \ldots$
For $n = 4$, we have that
$$\frac{x_5}{x_4} = 3\frac{1}{3}.$$
Thus, it follows that
$$3 < \frac{x_5}{x_4} < 4.$$
Assume that the statement holds true for $n = k$, where $k \geq 4$ That is
$$3 < \frac{x_{k+1}}{x_k} < 4.$$
Let us prove that the statement holds true for $n = k+1$. That is
$$3 < \frac{x_{k+2}}{x_{k+1}} < 4.$$
Note that
$$\frac{x_{k+2}}{x_{k+1}} = 5 - 5 \cdot \frac{x_k}{x_{k+1}} \in \left(\frac{10}{3}, \frac{15}{4}\right).$$
This ends the proof of the statement. Hence, the answer is 2.   □

Problem 4.300. Let a, b, c be integers, such that $a \neq 0$ and
$$3.9|a| \geq 2|b| + |c|.$$
Given that equation $ax^2 + bx + c = 0$ has only integer solutions. What is the total number of equations of the following form?
$$x^2 + \frac{b}{a}x + \frac{c}{a} = 0.$$

(A) 1   (B) 2   (C) 3   (D) 5   (E) 6

Solution. Answer. (E)
Let integer number k be a solution of the equation
$$ax^2 + bx + c = 0.$$
We have that
$$|ak^2| = |-bk - c| \leq |b||k| + |c|.$$
If $|k| \geq 2$, then
$$2|a| \leq |a||k| \leq |b| + \frac{|c|}{|k|} \leq |b| + \frac{|c|}{2} \leq \frac{3.9}{2}|a|.$$
Thus, it follows that
$$0.1|a| \leq 0.$$
This leads to a contradiction.
Let integers x_1 and x_2 be the solutions of the equation
$$x^2 + \frac{b}{a}x + \frac{c}{a} = 0.$$
Note that the following cases are possible:
$x_1 = -1, x_2 = -1$,
$x_1 = 0, x_2 = 0$,
$x_1 = 1, x_2 = 1$,
$x_1 = -1, x_2 = 0$,
$x_1 = -1, x_2 = 1$,
$x_1 = 0, x_2 = 1$.
The corresponding equations will be the following ones
$x^2 + 2x + 1 = 0$,
$x^2 = 0$,
$x^2 - 2x + 1 = 0$,
$x^2 + x = 0$,
$x^2 - 1 = 0$.
$x^2 - x = 0.$
Therefore, the answer is 6. $\square$

4.13 Solutions of AMC 12 type practice test 13

Problem 4.301. What is the value of the expression $\dfrac{2^{2020} + 2^{2021}}{4^{1011} - 4^{1010}}$?

(A) $\dfrac{1}{2}$ (B) 1   (C) 2   (D) 4   (E) 2^{16}

Solution. Answer. (B)
Note that
$$\frac{2^{2020} + 2^{2021}}{4^{1011} - 4^{1010}} = \frac{2^{2020}(1+2)}{4^{1010}(4-1)} = \frac{2^{2020}}{(2^2)^{1010}} = 1.$$

□

Problem 4.302. Let n be a positive integer, such that $n! = 240 \cdot (n-2)!$. What is the value of n?

(A) 13   (B) 14   (C) 15   (D) 16   (E) 17

Solution. Answer. (D)
Given that
$$n! = 240 \cdot (n-2)!.$$
Thus, it follows that
$$n(n-1) = 240.$$
Therefore $n = 16$.

□

Problem 4.303. *Given that the median of numbers 1, 2, x, 13 is equal to 3. What is the value of the mean of these numbers?*

(A) 5   (B) 6   (C) 10   (D) 11   (E) 12

Solution. Answer. (A)
Note that $1 \leq x \leq 13$. Given that the median of numbers 1, 2, x, 13 is equal to 3, then
$$\frac{2+x}{2} = 3.$$
Thus, it follows that $x = 4$.
The arithmetic mean of the numbers 1, 2, 4, 13 is equal to 5, as
$$\frac{1+2+4+13}{4} = 5.$$

□

Problem 4.304. *Let for positive real numbers x and y the operation $\star$ be defined in the following way $x \star y = \dfrac{1}{x} + \dfrac{1}{y}$. What is the value of the expression*
$$\sqrt{1} \star \sqrt{2} - \sqrt{2} \star \sqrt{3} + \sqrt{3} \star \sqrt{4} - \ldots - \sqrt{98} \star \sqrt{99} + \sqrt{99} \star \sqrt{100}?$$

(A) 0   (B) 0.1   (C) 1   (D) 1.01   (E) 1.1

254

Solution. Answer. (E)
Note that, according to the definition of operator $\star$, we have that

$$\sqrt{1}\star\sqrt{2}-\sqrt{2}\star\sqrt{3}+\sqrt{3}\star\sqrt{4}-...-\sqrt{98}\star\sqrt{99}+\sqrt{99}\star\sqrt{100} = \left(\frac{1}{\sqrt{1}}+\frac{1}{\sqrt{2}}\right)-\left(\frac{1}{\sqrt{2}}+\frac{1}{\sqrt{3}}\right)+\left(\frac{1}{\sqrt{3}}+\frac{1}{\sqrt{4}}\right)-...$$

$$-\left(\frac{1}{\sqrt{98}}+\frac{1}{\sqrt{99}}\right)+\left(\frac{1}{\sqrt{99}}+\frac{1}{\sqrt{100}}\right) = \frac{1}{\sqrt{1}}+\left(\frac{1}{\sqrt{2}}-\frac{1}{\sqrt{2}}\right)+\left(-\frac{1}{\sqrt{3}}+\frac{1}{\sqrt{3}}\right)+...+\left(-\frac{1}{\sqrt{99}}+\frac{1}{\sqrt{99}}\right)+\frac{1}{\sqrt{100}} =$$

$$= 1 + 0.1 = 1.1$$

□

Problem 4.305. *Given that 23 students are sitting in three rows, such that in each row there is at least one student. Given also that 20% of students of the first row, 25% of the second row and 10% of the third row are attending a basketball club. How many students from that class are attending a basketball club?*

(A) 3   (B) 4   (C) 5   (D) 6   (E) 7

Solution. Answer. (B)
Let m, n, p be the number of students of each row, respectively. According to the condition of the problem in these rows the number of students attending a basketball club is equal to $\frac{m \cdot 20}{100}, \frac{n \cdot 25}{100}, \frac{p \cdot 10}{100}$, respectively.
Thus, it follows that $5 \mid m$, $4 \mid n$ and $10 \mid p$.
We deduce that
$$m + n \geq 9.$$
Hence, we obtain that $p \leq 14$. Therefore
$$p = 10, m = 5, n = 8.$$
Thus, it follows that
$$\frac{m}{5} + \frac{n}{4} + \frac{p}{10} = 4.$$

□

Problem 4.306. *Given that the sum of three real numbers is 2 and their product is -2. Which of the following statements holds true?*

(A) All three numbers are positive.
(B) All three numbers are negative.
(C) One of the numbers is 0.
(D) Two of the numbers are negative and the third is positive.
(E) Two of the numbers are positive and the third is negative.

Solution. Answer. (E)
Let m, n, k be the given numbers.
Note that statements (A) and (D) do not hold true, as in these cases we obtain that $mnk > 0$.
Note that statement (B) does not hold true, as $m + n + k < 0$.
Note also that statement (C) does not hold true, as $mnk = 0$.
Finally, note that statement (E) is possible. For example, when $m = 2, n = 1, k = -1$.

□

Problem 4.307. *What is the greatest number of all three-digit numbers with equal sums of the digits?*

(A) 62   (B) 69   (C) 70   (D) 80   (E) 120

Solution. Answer. (C)
Let us consider the following table (see the figure).

x+y	0	1	2	3	4	5	6	7	8	9	10	11	12	13	14	15	16	17	18
The number of pairs of (x, y) digits	1	2	3	4	5	6	7	8	9	10	9	8	7	6	5	4	3	2	1

Taking this table into consideration, we have that the greatest number of three-digit numbers $\overline{zxy}$ with equal sums of the digits is equal to the sum of the numbers written in 9 consecutive cells of the table the sum of which is the greatest.

Thus, it follows that the greatest number of all such three-digit numbers is:
$$10 + 2 \cdot 9 + 2 \cdot 8 + 2 \cdot 7 + 2 \cdot 6 = 70.$$

Therefore, the total number of all such three-digit numbers is equal to 70 (and the sum of the digits for each of them is equal to 14). □

Problem 4.308. *Let BE be a median of triangle ABC and CF be a median of triangle BEC. Given that $AF = AE$ and $CF = 20$. What is the length of line segment AB?*

(A) 16   (B) 17   (C) 18   (D) 19   (E) 20

Solution. Answer. (E)
According to the assumption of the problem, we have that AEF is an isosceles triangle (see the figure).

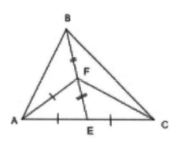

Thus, it follows that
$$\angle AEF = \angle AFE.$$
We obtain that
$$\angle AFB = 180° - \angle AFE = 180° - \angle AEF = \angle CEF.$$
We deduce that
$$AF = CE, BF = EF,$$
and
$$\angle AFB = \angle CEF.$$
Note that triangles AFB and CEF are congruent triangles. Thus, it follows that
$$AB = CF = 20.$$

□

Problem 4.309. *The factory did not work on any Saturdays or Sundays of February. On the nth day of each week it produced $(6-n)^2$ devices, where $n \in \{1,2,3,4,5\}$. Given that in total the factory produced 236 devices during the month of February. Which day of the week was the last day of that February?*

(A) Monday (B) Tuesday (C) Wednesday (D) Thursday (E) Friday

Solution. Answer. (B)
Note that during the first 28 days of February the factory produced $4(5^2 + 4^2 + 3^2 + 2^2 + 1^2) = 220$ devices.
Therefore, that February had 29 days and on the last day the factory produced $236 - 220 = 16$ devices.
Hence, the last day of that February was Tuesday. □

Problem 4.310. *60 scientists took part in a seminar. Given that 50 of them speak English and 45 of them speak French. Given also that each participant speaks at least one of these two languages. What is the probability that randomly chosen two scientists can communicate with each other either in English or in French?*

(A) $\dfrac{54}{59}$ (B) $\dfrac{5}{59}$ (C) $\dfrac{1}{2}$ (D) $\dfrac{2}{3}$ (E) $\dfrac{3}{5}$

Solution. Answer. (A)
The number of scientists participating in the seminar that speak in English but do not speak French is equal to 15, as $60 - 45 = 15$.
On the other hand, the number of scientists that speak French but do not speak English is equal to 10, as $60 - 50 = 10$.
Therefore, the probability that out of 60 scientists randomly selected 2 participants are not able to communicate with each other is
$$\frac{15 \cdot 10}{\binom{60}{2}} = \frac{15 \cdot 10}{30 \cdot 59} = \frac{5}{59}.$$

Thus, it follows that the probability that randomly chosen two scientists can communicate with each other either in English or in French is
$$1 - \frac{5}{59} = \frac{54}{59}.$$
□

Problem 4.311. *Let point M be inside the rhombus $ABCD$. Given that $MA = AB = \sqrt{2+\sqrt{2}}$, $\angle MAB = 15°$ and $\angle ADC = 105°$. What is the length of line segment MC?*

(A) 1 (B) $\sqrt{2}$ (C) $\sqrt{3}$ (D) 2 (E) $\sqrt{2+\sqrt{5}}$

Solution. Answer. (B)
Let us consider the following figure.

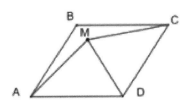

Note that
$$\angle BAD = 180° - 105° = 75°,$$
and
$$\angle MAD = \angle BAD - \angle MAB = 60°.$$

As $\angle MAD = 60°$ and $MA = AD$, then triangle MAD is an equilateral triangle. Thus, it follows that
$$MD = AD = CD,$$
and $\angle ADM = 60°$. Therefore $\angle MDC = 105° - 60° = 45°$.
According to the law of cosines from triangle MCD, we have that
$$MC^2 = CD^2 + CD^2 - 2 \cdot CD \cdot CD \cdot \cos 45° = (2 + \sqrt{2})(2 - \sqrt{2}) = 2.$$
Thus, it follows that $MC = \sqrt{2}$. □

Problem 4.312. *Positive integer n is called "amazing", if the sum of its two greatest divisors is equal to 42. What is the total number of all "amazing" numbers?*

(A) 4   (B) 3   (C) 5   (D) 2   (E) 6

Solution. Answer. (B)
Assume the smallest prime divisor of the "amazing" number n is p.
Note that, in this case the two greatest divisors of n are n and $\dfrac{n}{p}$.
According to the assumption of the problem, we have that the sum of two greatest divisors of n is equal to 42, therefore
$$n + \frac{n}{p} = 42.$$
Thus, it follows that
$$n = \frac{42p}{p+1}.$$
We deduce that
$$(p+1) \mid 42.$$
Taking this into consideration and as p is prime, we obtain that
$$p \in \{2, 5, 13, 41\}.$$
A straightforward verification shows that only the values $p = 2$, $p = 5$, $p = 41$ satisfy the assumptions of the problem and in these cases $n = 28$, $n = 35$, $n = 41$, respectively. Therefore, the total number of all "amazing" numbers is equal to 3. □

Problem 4.313. *What is the equation of the line that is symmetric to the line $y = 2x - 1$ with respect to line $y = x$?*

(A) $y = 2x + 1$ (B) $y = x - 0.5$ (C) $y = x + 0.5$ (D) $y = 0.5x + 0.5$ (E) $y = -x$

Solution. Answer. (D)
At first, let us find the coordinates of the intersection point A of the line $y = 2x - 1$ and the line $y = x$.
From the equation $2x - 1 = x$ we obtain that $x = 1$. Hence $y = 1$ and $A = (1, 1)$.
Note that point $B = (0, -1)$ belongs to the line $y = 2x - 1$. The reflection point of B with respect to the line $y = x$ is point $C = (-1, 0)$. We have to find the equation of the line AC.
Let the equation of the line AC be $y = kx + b$.
We have that $1 = k + b$ and $0 = -k + b$. Therefore $b = 0.5$ and $k = 0.5$.
Hence, we obtain that the equation of the line that is symmetric to the line $y = 2x - 1$ with respect to line $y = x$ is $y = 0.5x + 0.5$. □

Problem 4.314. Let $ABCD$ be a trapezoid with bases BC and AD, such that $BC = 20$ and $AD = 36$. Given that $AB = 63$ and $\angle ABC = 90°$. Let M be such a on leg AB that $BM = 15$. What is the value of the sum of the inradii of triangles ADM, BMC, CDM?

(A) 25   (B) 27   (C) 29   (D) 30   (E) 32

Solution. Answer. (B)
Let us denote by r_1, r_2, r_3 the inradii of triangles ADM, BMC, CDM, respectively (see the figure).

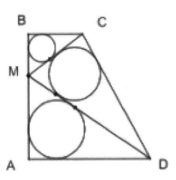

We have that
$$AM = AB - BM = 48.$$
Note that
$$\angle MAD = 90° = \angle CBM,$$
and
$$\frac{BM}{AD} = \frac{15}{36} = \frac{20}{48} = \frac{BC}{AM}.$$
Hence, we obtain that triangles BMC and ADM are similar. Therefore $\angle BCM = \angle AMD$. Thus, it follows that
$$\angle AMD + \angle BMC = \angle BCM + \angle BMC = 90°.$$
We deduce that
$$\angle CMD = 180° - (\angle BMC + \angle AMD) = 90°.$$
According to the Pythagorean theorem, we have that
$$CM^2 = 625, DM^2 = 3600, CD^2 = CM^2 + DM^2 = 65^2.$$
On the other hand, according to the formula of the radius of an inscribed circle of a right triangle, from triangles ADM, BMC, CDM, we obtain that
$$r_1 = \frac{AM + AD - DM}{2}, r_2 = \frac{BM + BC - CM}{2}, r_3 = \frac{CM + DM - CD}{2}.$$
Thus, it follows that
$$r_1 + r_2 + r_3 = \frac{AM + BM + AD + BC - CD}{2} = \frac{AB + AD + BC - CD}{2} = \frac{20 + 36 + 63 - 65}{2} = 27.$$
$\square$

Problem 4.315. *How many five-digit numbers ending with 12 are divisible by 13?*

(A) 23   (B) 27   (C) 60   (D) 68   (E) 69

Solution. Answer. (E)
Assume that a five-digit number $\overline{abc12}$ is divisible by 13. Note that

$$\overline{abc12} = \overline{abc00} + 12 = 100 \cdot \overline{abc} + 12 = 104 \cdot \overline{abc} - 4(\overline{abc} - 3) = 13 \cdot 8 \cdot \overline{abc} - 4(\overline{abc} - 3).$$

Thus, it follows that $13 \mid \overline{abc12}$ if and only if $13 \mid (\overline{abc} - 3)$. It is sufficient to find the number of all three-digit numbers which leave a remainder of 3 when divided by 13.
Note that these three-digit numbers form an arithmetic sequence with the first term equal to 107 and a common difference of 13.
From the following condition
$$107 + 13(n - 1) \leq 999,$$
we obtain that $n \in \{1, 2, ..., 69\}$. Therefore, there are 69 five-digit numbers satisfying the assumptions of the problem. $\square$

Problem 4.316. *Let M be a given point in triangle ABC, such that $\angle MCA = 60°$, $AB = 3MC$, $AC : MB = 2 : \sqrt{3}$. Given that $\angle BAC = 60°$. What is the measure (in degrees) of angle MBA?*

(A) 5   (B) 10   (C) 15   (D) 30   (E) 45

Solution. Answer. (D)
Let point N be the intersection of the lines CM and AB (see the figure).

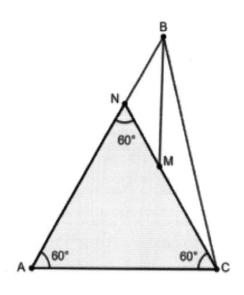

Given that
$$\angle NAC = 60°, \angle NCA = 60°.$$
Thus, it follows that $\angle ANC = 60°$.
Let $MC = x, MB = \sqrt{3}y$, according to the condition of the problem $AB = 3x$ and $AC = 2y$. From the equilateral triangle ANC, we get $AN = 2y$, $MN = 2y - x$. We have that $\angle BNM = 120°$ and $BN = 3x - 2y$. From triangle BMN according to the law of cosines, we obtain that

$$3y^2 = (2y - x)^2 + (3x - 2y)^2 - 2(2y - x)(3x - 2y)\cos 120°.$$

Thus, it follows that
$$y^2 - 8xy + 7x^2 = 0.$$
Hence, we deduce that either $y = x$ or $y = 7x$.
Note that the case $y = 7x$ does not hold true, as $3x > 2y$.
Therefore $y = x$. Hence $BN = 3x - 2y = x$ and $MN = 2y - x = x$. Thus, by the law of cosines, we obtain that $\angle MBA = 30°$. □

Problem 4.317. *What is the greatest number of pairwise distinct positive integers such that the positive difference of each two of them is a prime number?*

(A) 3   (B) 4   (C) 5   (D) 6   (E) 10

Solution. Answer. (B)
Note that such five numbers do not exist, otherwise according to the Pigeonhole Principle at least three of them have the same parity. On the other hand, the difference of the greatest and the smallest numbers is an even number which is not less than 4. Hence, this difference cannot be a prime number. Therefore, such five numbers do not exist.
Note that, an example of such four numbers is: 2, 4, 7, 9.
Therefore, the greatest number of pairwise distinct positive integers such that the positive difference of each two of them is a prime number is 4. □

Problem 4.318. *Consider all real values of a such that the equation $4^{x^2-5x} - 2^{x^2-5x} + a = 0$ has real solutions. For each of these values of a let $S(a)$ be sum of all real solutions of given equation. What is the greatest possible value of $S(a)$?*

(A) 2   (B) 4   (C) 5   (D) 10   (E) 25

Solution. Answer. (D)
Denote
$$y = 2^{x^2-5x}.$$
Then, we obtain that
$$y^2 - y + a = 0.$$
If this equation has a positive root y_0, then we obtain the following equation with respect to variable x
$$x^2 - 5x - \log_2 y_0 = 0.$$
Taking this into consideration, according to Vieta's formula, we obtain that
$$S(a) \in \{2.5, 5, 7.5, 10\}.$$
Note that, if $a = \dfrac{2}{9}$, then $S(a) = 10$.
Therefore, the greatest possible value of $S(a)$ is equal to 10. □

Problem 4.319. *Let $ABCD$ be a square with a side length of $a = 2\sqrt{1+\sqrt{3}}$. Consider the arcs of the circles with centers A, B, C, D and with radius $\dfrac{a}{2}$ which do not contain points outside the square (see the figure). Each of those 4 arcs are divided into 3 equal parts with 2 points. What is the area of the octagon with vertices at the above mentioned eight points?*

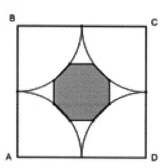

(A) $\dfrac{\sqrt{3}}{4}$ (B) $\dfrac{\sqrt{3}}{2}$ (C) 1   (D) 2   (E) 3

Solution. Answer. (D)
Note that the area S of the octagon with vertices of the above mentioned eight points (see the figure) is equal to $a^2 - 12 \cdot Area(BEF) - 4 \cdot Area(EFK)$.

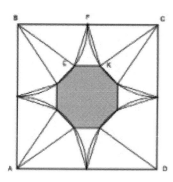

We have that $\angle EBF = 30°$. Therefore $\angle EFB = \angle KFC = 75°$ and $\angle EFK = 30°$. According to the law of cosines, from triangle BEF, we obtain that

$$EF^2 = \left(\frac{a}{2}\right)^2 + \left(\frac{a}{2}\right)^2 - 2 \cdot \frac{a}{2} \cdot \frac{a}{2} \cos 30° = \frac{a^2}{4}(2 - \sqrt{3}).$$

We have that
$$Area(BEF) = \frac{1}{2} \cdot \frac{a}{2} \cdot \frac{a}{2} \cdot \sin 30° = \frac{a^2}{16},$$

and
$$Area(EFK) = \frac{1}{2} EF^2 \cdot \sin 30° = \frac{a^2}{16}(2 - \sqrt{3}).$$

Thus, it follows that
$$S = a^2 - 12 \cdot \frac{a^2}{16} - 4 \cdot \frac{a^2}{16}(2 - \sqrt{3}) = \frac{a^2}{4}(\sqrt{3} - 1) = (\sqrt{3} + 1)(\sqrt{3} - 1) = 2.$$

□

Problem 4.320. A three-digit number $\overline{abc}$ written using some of the digits 0, 1, 2, 3, 4 is randomly chosen. What is the probability that the inequalities $|a - b| \geq 2$ and $|b - c| \geq 2$ simultaneously hold true?

(A) 0.05   (B) 0.1   (C) 0.2   (D) 0.23   (E) 0.25

Solution. Answer. (D)
Let $i \to j$ means that in three-digit number $\overline{abc}$ in the right side of the digit i is written digit j.
All three-digit numbers $\overline{abc}$ satisfying the inequalities $|a - b| \geq 2$ and $|b - c| \geq 2$ can be shown as follows (see the figure).

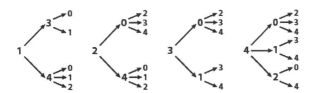

Therefore, the total number of all possible outcomes is equal to $4 \cdot 5 \cdot 5 = 100$ and the total number of all favorable outcomes is equal to 23.
Thus, the probability that the given inequalities simultaneously hold true is equal to $\dfrac{23}{100} = 0.23$. □

Problem 4.321. Let C_1 and B_1 be tangent points of the inscribed circle of triangle ABC with the sides AB and AC, respectively. Given that $\angle ACB = 92°$. Points M and N lie on the side AB, such that $C_1M = C_1N = CB_1$. What is the measure (in degrees) of angle MCN?

(A) 46   (B) 48   (C) 45   (D) 50   (E) 44

Solution. Answer. (E)
Let us consider point K on side AC, such that line segment MK is tangent to the given circle (see the figure).

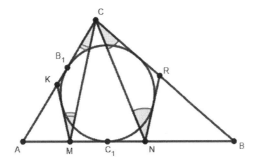

We have that $CB_1 = MC_1$, hence
$$\angle C_1MK = \angle ACB = 92°.$$
On the other hand, we have that $CK = KM$. Thus $\angle KCM = \angle KMC = \alpha$.
In a similar way, we obtain that $\angle C_1NR = 92°$ and $RC = NR$.
Hence, we deduce that $\angle CNR = \angle NCR = \beta$.
Note that, from pentagon $MKCRN$ we have that
$$3 \cdot 92° + 180° - 2\alpha + 180° - 2\beta = 540°.$$

Thus, it follows that $\alpha + \beta = 48°$. Therefore
$$\angle MCN = 92° - 48° = 44°.$$
$\square$

Problem 4.322. *What is the sum of all positive integers n less than 100, such that for each of them the equality $\frac{1}{n} + \frac{1}{n+2} = 0.\overline{a_n b_n c_n d_n e_n f_n}$ holds true, where $a_n, b_n, c_n, d_n, e_n, f_n$ are some digits?*

(A) 61 (B) 69 (C) 63 (D) 65 (E) 67

Solution. Answer. (D)
According to the condition of the problem, we have that
$$\frac{2(n+1)}{n(n+2)} = \frac{\overline{a_n b_n c_n d_n e_n f_n}}{999999}.$$
Therefore, number $2 \cdot 999999 \cdot n(n+1)$ is divisible by $n(n+2)$. Hence, n is odd and $n+1$ is relatively prime to each of the numbers n and $n+2$. Thus, it follows that
$$n(n+2) \mid 999999 = 3 \cdot 3 \cdot 7 \cdot 11 \cdot 13 \cdot 37.$$
A straightforward verification shows that $n < 100$ and the last equation holds true, then $n \in \{1, 7, 9, 11, 37\}$. Therefore, the sum of all positive integers n less than 100, such that for each of them the given inequality holds true is $1 + 7 + 9 + 11 + 37 = 65$.
$\square$

Problem 4.323. *What is the sum of the greatest and smallest solutions of the equation $\cos x - \sin x + \cos 4x = -0.5$ in the interval $\left[0, \frac{5\pi}{3}\right]$?*

(A) $\frac{2\pi}{3}$ (B) π (C) $\frac{4\pi}{3}$ (D) $\frac{3\pi}{2}$ (E) $\frac{5\pi}{3}$

Solution. Answer. (D)
Note that the given equation does not have any solutions in the interval $\left[\frac{3\pi}{2}, \frac{5\pi}{3}\right]$.
Let us consider $x \in \left[\frac{3\pi}{2}, \frac{5\pi}{3}\right]$, then we have that $4x \in \left[6\pi, \frac{20\pi}{3}\right]$. Thus, it follows that $\cos x \geq -0.5$.
Besides that, we have that
$$\cos x - \sin x > 0, (\cos x \geq 0 > \sin x).$$
Adding up the last two inequalities, we obtain that
$$\cos x - \sin x + \cos 4x > -0.5.$$
This ends the proof of the statement.
According to this statement, we also have that
$$\cos\left(\frac{3\pi}{2} - x\right) - \sin\left(\frac{3\pi}{2} - x\right) + \cos 4\left(\frac{3\pi}{2} - x\right) = -\sin x - (-\cos x) + \cos 4x = \cos x - \sin x + \cos 4x.$$
Hence, we deduce that if x_0 is a solution of the given equation, then $\frac{3\pi}{2} - x_0$ is a solution too.
Taking this into consideration, we obtain that if x_1 is the smallest solution of the given equation in the interval $\left[0, \frac{5\pi}{3}\right]$, then $x_1 \neq \frac{3\pi}{2} - x_1$ and $\frac{3\pi}{2} - x_1$ is the greatest solution in the interval $\left[0, \frac{5\pi}{3}\right]$.
Therefore, the sum of the greatest and smallest solutions of the given equation is equal to $\frac{3\pi}{2}$, as
$$x_1 + \left(\frac{3\pi}{2} - x_1\right) = \frac{3\pi}{2}.$$
$\square$

Problem 4.324. *Let $\lambda(n)$ be the number of all divisors of n. Given that first $f(n)$ factors form an arithmetic sequence, but the first $f(n) + 1$ factors do not. What is the greatest possible value of $2f(n) - \lambda(n)$?*

(A) -1 (B) 0 (C) 1 (D) 2 (E) 7

Solution. Answer. (D)
Note that
$$f(6) = 3, \lambda(6) = 4,$$
and
$$2f(6) - \lambda(6) = 2.$$
Let us prove that for some positive integer n the following inequality holds true
$$2f(n) - \lambda(n) \leq 2.$$
Let m be a positive integer, such that $2f(m) - \lambda(m) \geq 3$.
Let the first $k = f(m)$ factors of m be the numbers $d_1 < d_2 < ... < d_k$.
According to our assumption, we have that
$$\lambda(m) - k \leq k - 3.$$
Therefore, there is a numbers s, such that $k - s > 1$ and $d_s \cdot d_k = n$.
According to the assumption of the problem, we have that
$$d_k - d_{k-1} = \frac{n}{d_{k-1}} - \frac{n}{d_k}.$$
Note that this leads to a contradiction.
Remark. We use that
$$\{d_1, d_2, ..., d_{\lambda(n)}\} = \{\frac{n}{d_{\lambda(n)}}, ..., \frac{n}{d_2}, \frac{n}{d_1}\}.$$
$\square$

Problem 4.325. *Let set $S = \{(i, j) \mid i \in \{1, 2, ..., 10\}$ and $j \in \{1, 2, ..., 10\}\}$ is given on the coordinate plane. Triangle ABC is chosen on the plane, such that none of the points A, B, C belong to S. Let n be the number of points in set S, such that each of them lies on one of the sides of triangle ABC. What is the greatest possible value of n?*

(A) 10 (B) 20 (C) 22 (D) 23 (E) 30

Solution. Answer. (B)
Let l_i be the line $x = i$, where $i = 1, 2, ..., 10$.
When the sides of triangle ABC do not lie on any of those lines, then each of the lines l_i cannot have more than 2 common points with the sides of triangle ABC. Besides, each element of the set S belongs to any line l_i. Thus, it follows that $n \leq 20$.
When a side of triangle ABC lies on any of those lines, then we consider lines $y = i$, where $i = 1, ..., 10$ and in a similar way we obtain that $n \leq 20$.
Let us provide an example where $n = 20$ (see the figure).

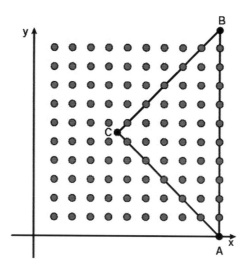

Therefore, the greatest possible value of n is 20. □

4.14 Solutions of AMC 12 type practice test 14

Problem 4.326. *What is the value of the expression $(15^{-2} + 20^{-2})^{-1} - (15^{-1} - 20^{-1})^{-1}$?*

(A) 144   (B) 100   (C) 84   (D) 48   (E) 4

Solution. Answer. (C)
Note that

$$(15^{-2}+20^{-2})^{-1}-(15^{-1}-20^{-1})^{-1} = (5^{-2})^{-1}(3^{-2}+4^{-2})^{-1}-(5^{-1})^{-1}(3^{-1}-4^{-1})^{-1} = 25 \cdot \frac{144}{25} - 5 \cdot 12 = 84.$$

$\square$

Problem 4.327. *David and Anna ate 7 candies altogether. After David ate another 5 candies, the number of candies that he ate was twice the number of candies that Anna ate. How many candies did Anna eat?*

(A) 3   (B) 4   (C) 5   (D) 6   (E) 2

Solution. Answer. (B)
Note that eventually they ate altogether 12 candies ($7 + 5 = 12$).
According to the assumption of the problem, Anna ate $\frac{1}{3}$ of the total number of candies.
Therefore, Anna ate 4 candies ($12 \cdot \frac{1}{3} = 4$).

$\square$

Problem 4.328. *What is the value of the sum $0.10 + 0.11 + ... + 0.99$?*

(A) 49   (B) 49.01   (C) 49.03   (D) 49.05   (E) 49.5

Solution. Answer. (D)
Note that

$$0.10 + 0.11 + ... + 0.99 = (10 + 11 + ... + 99) \cdot 0.01 = \frac{10 + 99}{2} \cdot 90 \cdot 0.01 = 49.05.$$

$\square$

Problem 4.329. *What is the sum of all positive integers n such that $n!$ ends with exactly 6 zeros?*

(A) 135   (B) 107   (C) 78   (D) 51   (E) 25

Solution. Answer. (A)
According to the condition of the problem, we have that the power of 5 in the prime factorization of $n!$ should be 6. Thus, it follows that

$$25 \leq n < 30.$$

Therefore, the sum of all positive integers n such that $n!$ ends with exactly 6 zeros is $25+26+27+28+29 = 135$.

$\square$

Problem 4.330. *A two-digit number is 8 times greater than the sum of its digits. What is the value of the sum of the digits of this two-digit number?*

(A) 5   (B) 7   (C) 9   (D) 11   (E) 12

Solution. Answer. (C)
Let $\overline{ab}$ be the given two-digit number. According to the assumption of the problem, we have that
$$\overline{ab} = 8(a+b).$$
Thus, it follows that
$$10a + b = 8a + 8b.$$
Hence, we obtain that $2a = 7b$. Therefore $a = 7, b = 2$. Thus, it follows that
$$a + b = 9.$$

□

Problem 4.331. *Given that the sum of the ages of all students in the classroom is 120. After 6 years the sum of their ages will be twice as much as the sum of their ages 3 years ago. What is the value of the arithmetic mean of the ages of all students?*

(A) 8   (B) 10   (C) 12   (D) 15   (E) 16

Solution. Answer. (C)
Let n be the number of students in the class. After 6 years the sum of their ages will be $120 + 6n$, and 3 years ago the sum of their ages was $120 - 3n$.
Therefore, according to the assumption of the problem, we obtain that $120 + 6n = 2(120 - 3n)$.
Thus, it follows that $n = 10$. Hence, the arithmetic mean of their ages is equal to 12, as $\dfrac{120}{10} = 12$. □

Problem 4.332. *In a geometric sequence with positive terms, the difference between the seventh and the first terms is seven times more than the sum of the fourth and the first terms. What is the common ratio of this geometric sequence?*

(A) 1   (B) $\sqrt[3]{2}$ (C) 2   (D) 3   (E) 3.2

Solution. Answer. (C)
Let (b_n) be the given geometric sequence. We have that
$$b_7 - b_1 = 7(b_1 + b_4).$$
Thus, it follows that
$$b_1(r^6 - 1) = 7b_1(1 + r^3),$$
where $r = \dfrac{b_2}{b_1} > 0$.
Hence, we obtain that $r^3 - 1 = 7$.
Therefore $r = 2$.

□

Problem 4.333. *David solves five problems each Saturday and Sunday, and he solves six problems each weekday. During some consecutive days, David solved 70 problems. Which day of the week did he start solving the problems?*

(A) Wednesday (B) Thursday (C) Friday (D) Sunday (E) Monday

Solution. Answer. (E)
Note that $6 \mid (70 - 2 \cdot 5)$ and $6 \mid (70 - 8 \cdot 5)$.
On the other hand, if during some consecutive days the number of Saturdays and Sundays is 8, then the number of weekdays is 5, which is impossible.
Therefore, the number of Saturdays and Sundays is 2. Hence, the number of weekdays is 10. □

Problem 4.334. *Let $ABCD$ be a square with a side length $\sqrt{3}$. Let square $ABCD$ be rotated around point A by $30°$ and as a result we obtained square $AB'C'D'$. What is the area of the part that squares $AB'C'D'$ and $ABCD$ have in common?*

(A) 1 (B) 1.5 (C) $\sqrt{3}$ (D) 2 (E) 2.1

Solution. Answer. (C)
Consider the following figure.

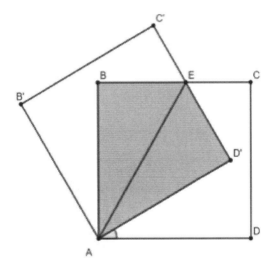

Note that right triangles ABE and AED' are congruent, therefore

$$\angle BAE = \angle D'AE = \frac{60°}{2} = 30°.$$

Hence, from triangle ABE we obtain that $BE = 1$ and

$$Area(ABE) = \frac{\sqrt{3}}{2}.$$

Thus, it follows that

$$Area(ABED') = 2\,Area(ABE) = \sqrt{3}.$$

□

Problem 4.335. *Let us choose any random moment between 12 AM and 12 PM. What is the probability of the event that all three arrows of the clock are located on the right semi-circle of the clock?*

(A) $\dfrac{1}{2}$ (B) $\dfrac{1}{3}$ (C) $\dfrac{1}{8}$ (D) $\dfrac{1}{12}$ (E) $\dfrac{1}{15}$

Solution. Answer. (C)
The probability of the event that any of these arrows will be on the right semi-circle of the clock during 12 hours is $\dfrac{1}{2}$.
Therefore, the probability of the event that all the three arrows will be in the right semi-circle of the clock is $\dfrac{1}{2} \cdot \dfrac{1}{2} \cdot \dfrac{1}{2} = \dfrac{1}{8}$. □

Problem 4.336. *Given that $2\sin x + \cos x = \sqrt{5}$. What is the value of $\tan x$?*

(A) $\sqrt{2}$ (B) 2 (C) $\sqrt{3}$ (D) 1 (E) $\dfrac{\sqrt{5}}{5}$

Solution. Answer. (B)
Note that
$$(2\sin x + \cos x)^2 + (2\cos x - \sin x)^2 = 4\sin^2 x + 4\sin x\cos x + \cos^2 x + 4\cos^2 x - 4\cos x\sin x + \sin^2 x =$$
$$= 5(\sin^2 x + \cos^2 x) = 5.$$

Thus, it follows that
$$(2\sin x + \cos x)^2 + (2\cos x - \sin x)^2 = 5.$$

On the other hand, we have that
$$2\sin x + \cos x = \sqrt{5}.$$

Hence, we obtain that
$$2\cos x - \sin x = 0.$$

Therefore $\tan x = 2$.
Alternative solution. Apply Cauchy-Bunyakovsky-Schwarz inequality, then
$5 = (2\sin x + \cos x)^2 \leq (2^2 + 1^2)(\sin^2 x + \cos^2 x) = 5$, where the equality holds true if and only if $\dfrac{\sin x}{2} = \dfrac{\cos x}{1}$. Note that $\cos x \neq 0$ (otherwise $\cos x = 0$ and $\sin x = 0$), then $\tan x = 2$. □

Problem 4.337. *Three sisters bought 4 identical bracelets. In how many different ways can they wear those 4 bracelets? (they can wear each bracelet either on right or left hand).*

(A) 15 (B) 20 (C) 24 (D) 120 (E) 126

Solution. Answer. (E)
Assume the first sister wears x_1 bracelet(s) on her left hand and x_2 bracelet(s) on the right hand.
In the same way we define numbers x_3, x_4, x_5, x_6. We need to find the total number of non-negative integer solutions of the following equation
$$x_1 + x_2 + x_3 + x_4 + x_5 + x_6 = 4.$$

The total number of non-negative integer solutions of this equation is equal to
$$\binom{9}{5} = 126.$$

□

Problem 4.338. *Let ABC be a triangle, such that $AB = 12$, $BC = 9$, $AC = 16$. Given that point D lies on the side AC and point E lies on the ray DB, so that $CD = 7$ and $\angle DAE = \angle ABD$. What is the length of line segment AE?*

(A) 9 (B) 11 (C) 12 (D) 16 (E) 18

Solution. Answer. (D)
We have that $AD = AC - CD = 9$.

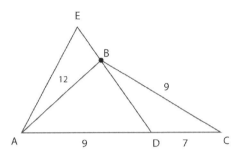

Note that
$$\frac{AD}{AB} = \frac{9}{12} = \frac{12}{16} = \frac{AB}{AC}.$$

Thus, it follows that triangles ADB and ABC are similar. Therefore
$$\angle DAE = \angle ABD = \angle ACB.$$

Hence, we obtain that $AD = BC$ and
$$\angle DAE = \angle ACB, \angle ADB = \angle ABC.$$

Therefore, triangles ADE and CBA are congruent triangles.
Thus, it follows that
$$AE = AC = 16.$$

$\square$

Problem 4.339. *Let p be a prime number, such that $4p - 113$ and $4p + 113$ are prime numbers. What is the value of the sum of all the digits of p?*

(A) 10 (B) 11 (C) 21 (D) 25 (E) 30

Solution. Answer. (B)
Note that
$$4p - 113 - (p+1) = 3(p - 38).$$
Thus, it follows that $4p - 113$ and $p + 1$ leave the same remainders when divided by 3.
In a similar way, one can prove that $4p + 113$ and $p + 2$ leave the same remainders when divided by 3.
Therefore, numbers p, $4p - 113$, $4p + 113$ leave respectively the same remainders as numbers p, $p + 1$, $p + 2$ when divided by 3.
Hence, one of the numbers p, $4p - 113$, $4p + 113$ is divisible by 3. On the other hand, as these numbers are prime, then one of them is equal to 3.
Note that, among these three numbers only $4p - 113$ can be equal to 3.
We have that $4p - 113 = 3$. Therefore $p = 29$.
Note also that the case $p = 29$ satisfies the assumptions of the problem, as we have that $4p - 113 = 3$, $4p + 113 = 229$, which are prime numbers.

$\square$

Problem 4.340. *Let $ABCD$ be a square with a side lenght of $2\sqrt{5}$. Let M, N, P, K be the midpoints of sides AB, BC, CD, AD, respectively. What is the radius of the circle that is tangent to each of line segments AN, DN, AP, BP, BK, CK, CM, DM?*

(A) $\dfrac{\sqrt{5}}{2}$ (B) 1 (C) $\dfrac{\sqrt{3}}{2}$ (D) $\dfrac{\sqrt{2}}{2}$ (E) $\dfrac{1}{2}$

Solution. Answer. (B)
Consider the following figure.

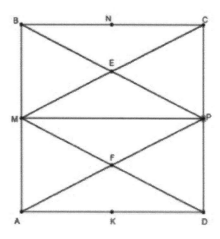

Note that we need to find the radius of the circle inscribed in the rhombus $MEPF$. We have that

$$Area(MEPF) = \frac{1}{4} Area(ABCD) = 5,$$

and

$$ME = \frac{1}{2} MC = \frac{1}{2}\sqrt{MB^2 + BC^2} = \frac{5}{2}.$$

Thus, it follows that

$$r = \frac{Area(MEPF)}{\frac{1}{2}(ME + EP + PF + MF)} = 1.$$

□

Problem 4.341. *At least 11 mathematicians took part in the Math conference. Given that each participant is acquainted with no more than 11 participants. Given also that for each group of 10 participants (of the conference) there exists a participant in the conference who is acquainted with all these 10 participants. What is the greatest number of the participants in this Math conference?*

(A) 11 (B) 12 (C) 13 (D) 14 (E) 15

Solution. Answer. (B)
Let n be the number of the participants in this Math conference.
According to the condition of the problem, we have that $n \geq 11$.
On the other hand, for each 10-member group there is a participant who is acquainted with these 10 people.
Obviously, for each participant there are at most 11 such 10-member groups (as each participant is acquainted with at most 11 people). Thus, it follows that

$$n \geq \frac{\binom{n}{10}}{11}.$$

Hence, we obtain that
$$\frac{(n-1)!}{11!(n-10)!} \leq 1.$$

We deduce that, either $n = 11$ or $n = 12$.

As we are looking for the greatest possible number of the participants, then it is sufficient to provide an example for $n = 12$. For example, if there are 12 participants and of 2 participants are acquainted with each other, then the conditions of the problem are satisfied. □

Problem 4.342. *A circle with center I is inscribed in the pentagon $ABCDE$. Given that $\angle BCD = 100°$. Let $\angle AIB + \angle EID = n°$. What is the value of n?*

(A) 100   (B) 120   (C) 140   (D) 150   (E) 160

Solution. Answer. (C)

Note that point I is equidistant from the sides of $\angle ABC$ and $\angle BAE$ (see the figure). Thus, it is on the bisector of each of them.

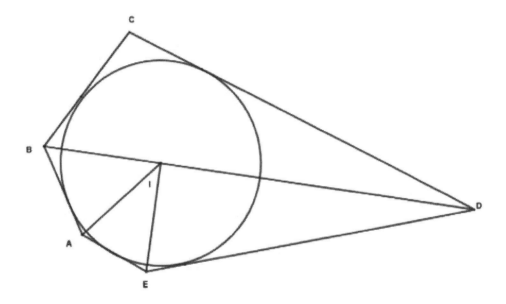

From triangle ABI, we have that
$$\angle AIB = 180° - \frac{1}{2}\angle ABC - \frac{1}{2}\angle BAE.$$

In a similar way, we obtain that
$$\angle EID = 180° - \frac{1}{2}\angle CDE - \frac{1}{2}\angle AED.$$

Summing up the last two equations, we deduce that
$$\angle AIB + \angle EID = 360° - \frac{1}{2}(\angle ABC + \angle BAE + \angle CDE + \angle AED) =$$
$$360° - \frac{1}{2}(540° - 100°) = 140°.$$

Thus, it follows that $n = 140$. □

Problem 4.343. *Each side of a convex hexagon ABCDEF is colored either white, blue or red. Given that each two sides with common vertex have different colors. In how many different ways is it possible to color all sides of this hexagon?*

(A) 36   (B) 48   (C) 66   (D) 97   (E) 98

Solution. Answer. (C)
Let us consider sides AB, CD and EF of $ABCDEF$.
Consider the following three cases:
Case 1. Sides AB, CD and EF are colored in different colors.
Then, the remaining sides are colored in a unique way.
In this case, all sides of the hexagon can be colored in 6 ways ($3 \cdot 2 \cdot 1 = 6$).
Case 2. Sides AB, CD and EF are colored in 2 different colors.
Then, the other 2 sides can be colored in 2 ways.
In this case all sides of the hexagon can be colored in 36 ways ($3 \cdot 3 \cdot 2 \cdot 2 = 36$).
Case 3. Sides AB, CD and EF are colored in the same color.
Then, the other sides can be colored in 8 ways ($2 \cdot 2 \cdot 2 = 8$). Hence, all sides of the hexagon can be colored in 24 ways ($3 \cdot 8 = 24$).
Thus, it follows that all sides of the hexagon is possible to color in 66 ways ($6 + 36 + 24 = 66$). □

Problem 4.344. *Let ABC be a triangle, such that $AB < BC$. Let D be the midpoint of the arc AC of the circumcircle of ABC, such that points B and D are on the different sides of line AC. Let line segment DE is perpendicular to the chord BC and point E belongs to chord BC. Given that $BE = 17$ and $EC = 7$. What is the length of side AB?*

(A) 7   (B) 8   (C) 9   (D) 10   (E) 11

Solution. Answer. (D)
Given that $\angle ABD = \angle CBD$ (see the figure).

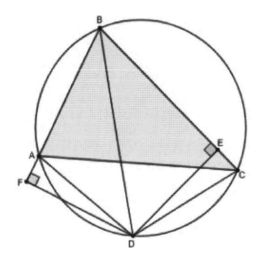

Let $DF \perp AB$, then $DE = DF$. On the other hand $\angle DAF = \angle DCE$. Thus, it follows that triangles DAF and DCE are congruent.
Taking this into consideration, we obtain that
$$AF = CE = 7.$$
On the other hand, triangles BED and BFD are congruent. Hence, we obtain that
$$BF = BE = 17.$$

We deduce that
$$AB = BF - AF = 17 - 7 = 10.$$
□

Problem 4.345. *For which value of a the equation $\log_a(x^2 - 4x + a + 4) = \log_{a^2-3a+7}(8x - 2x^2 - 3)$ has only one solution?*

(A) 6 (B) 5 (C) 4 (D) 3 (E) 2

Solution. Answer. (E)
Let us rewrite the given equation in the following way.
$$\log_a((x-2)^2 + a) = \log_{a^2-3a+7}(5 - 2(x-2)^2).$$
Note that if x_0 is a solution of the above equation, then $4 - x_0$ is also a solution of this equation. Therefore, the given equation has a unique solution if $4 - x_0 = x_0$. Hence $x_0 = 2$.
Now, let us check for which value of a we have that 2 is a solution of the given equation? We have that
$$\log_a a = \log_{a^2-3a+7} 5.$$
Thus, it follows that
$$a^2 - 3a + 7 = 5.$$
Hence, we obtain that either $a = 1$ or $a = 2$. Note that a is the base of the logarithm, so the case $a = 1$ does not satisfy the assumptions of the problem.
Let us verify that when $a = 2$, then the given equation has only one solution.
$$\log_2((x-2)^2 + 2) = \log_5(5 - 2(x-2)^2) = t.$$
Thus, it follows that
$$(x-2)^2 = 2^t - 2,$$
and
$$5 - 2(x-2)^2 = 5^t.$$
We deduce that
$$5 - 2(2^t - 2) = 5^t.$$
Hence, we obtain that
$$9 = 5^t + 2 \cdot 2^t.$$
Note that the function $f(t) = 5^t + 2 \cdot 2^t$ is increasing. Thus, the only solution of the last equation is $t = 1$. Therefore $x = 2$ is the only solution of the given equation. □

Problem 4.346. *What is the sum of all three-digit numbers, such that for each of them the sum of its digits is divisible by 7?*

(A) 2000 (B) 3500 (C) 50000 (D) 69237 (E) 75730

Solution. Answer. (D)
Let M be the set of all three-digit numbers, such that for each of them the sum of its digits is divisible by 7. Note that if $\overline{abc} \in M$, then $\overline{a_1 b_1 c_1} \in M$, where $a_1 = 10 - a$, $b_1 = 9 - b$, $c_1 = 9 - c$.
Note that if $\overline{abc} \in M$, then $7 \mid (a + b + c)$. Therefore
$$a_1 + b_1 + c_1 = 28 - (a + b + c),$$
and the last expression is divisible by 7. Note that $\overline{abc} \neq \overline{a_1 b_1 c_1}$ as b and b_1 have different parities. We call these three-digit numbers $\overline{abc}$ and $\overline{a_1 b_1 c_1}$ "pairs".

Let us assume that the total number of all "pairs" is equal to n. Then, the sum of all the elements of the set M is equal to $1099 \cdot n$.

Now, let us find n which is 5 less than the total number of all these three-digit numbers $\overline{abc}$, such that for each of them either $a+b+c = 7$ or $a+b+c = 14$. Moreover, when $a+b+c = 14$, then we have that $a \leq 5$.

Thus, it follows that
$$n = (7 + 6 + ... + 1) + (10 + 9 + 8 + 7 + 6) - 5 = 63.$$

Therefore, the sum of all three-digit numbers satisfying the assumptions of the problem is equal to 69237 ($1099 \cdot 63 = 69237$). $\square$

Problem 4.347. *Let AD be an angle bisector of triangle ABC. Given that $BD = 7$, $CD = 8$, $AC - AB = 13$ and $\angle A + 2\angle C = n°$. What is the value of n?*

(A) 60   (B) 75   (C) 90   (D) 105   (E) 120

Solution. Answer. (A)

Let us take point E on side AC, such that $AE = AB$ (see the figure).

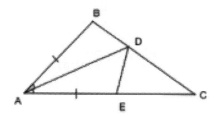

Note that triangles ABD and AED are congruent triangles. Therefore $ED = BD = 7$ and $\angle ADB = \angle ADE$. We have that
$$EC = AC - AE = AC - AB = 13.$$

From triangle DEC, according to the law of cosines, we obtain that
$$EC^2 = ED^2 + CD^2 - 2ED \cdot CD \cos \angle EDC.$$

Thus, it follows that
$$\cos \angle EDC = \frac{7^2 + 8^2 - 13^2}{2 \cdot 7 \cdot 8} = -\frac{1}{2}.$$

We deduce that $\angle EDC = 120°$. Therefore $\angle ADB = 30°$. We have that
$$\angle ADB = \frac{1}{2}\angle A + \angle C.$$

Thus, it follows that
$$n° = \angle A + 2\angle C = 60°.$$

Hence, we obtain that $n = 60$. $\square$

Problem 4.348. *What is the sum of all digits of the greatest integer, which is not possible to express as a sum of two positive integers, such that each of them is divisible by a square of an integer greater than 1?*

(A) 2   (B) 4   (C) 5   (D) 7   (E) 11

Solution. Answer. (C)
Note that
$$4n = 4 + 4(n-1),$$
and
$$4n + 2 = 18 + 4(n-4).$$
This means that any even number greater than 20 can be expressed as a sum of 2 integers, such that each of them is divisible by a square of an integer greater than 1.
Let us call such numbers "interesting numbers".
Note also that
$$4n + 1 = 9 + 4(n-2),$$
$$4n + 3 = 27 + 4(n-6),$$
and $27 = 9 + 18$. Therefore any odd number greater than 23 is an "interesting number".
A straightforward verification shows that 23 is not an "interesting number". Therefore, the sum of the digits of the greatest integer satisfying the assumptions of the problem is equal to 5 $(2 + 3 = 5)$. $\square$

Problem 4.349. *Let M be the greatest possible value of the expression $|\sin x(1-\cos y)| + |\sin y(1-\sin x)|$. What is the value of $(M-1)^2$?*

(A) 4   (B) 5   (C) 6   (D) 6.1   (E) 6.2

Solution. Answer. (B)
Note that
$$|\sin x(1-\cos y)| + |\sin y(1-\sin x)| = |\sin x|(1-\cos y) + |\sin y|(1-\sin x) =$$
$$= |\sin x| + |\sin y|(1-\sin x) - \cos y |\sin x|.$$
According to the Cauchy-Bunyakovsky-Schwarz inequality, we have that
$$|\sin y|(1-\sin x) - \cos y |\sin x| \le \sqrt{\left(|\sin y|^2 + (-\cos y)^2\right)\left((1-\sin x)^2 + \sin^2 x\right)} =$$
$$= \sqrt{1 - 2\sin x + 2\sin^2 x}.$$
Thus, it follows that
$$|\sin x(1-\cos y)| + |\sin y(1-\sin x)| \le |\sin x| + \sqrt{1 - 2\sin x + 2\sin^2 x} \le 1 + \sqrt{5}.$$
Note that, when $x = -\dfrac{\pi}{2}$, $y = \dfrac{\pi}{2} + \arccos \dfrac{2}{\sqrt{5}}$, then
$$|\sin x(1-\cos y)| + |\sin y(1-\sin x)| = 1 - \cos y + 2 \sin y =$$
$$= 1 + \dfrac{1}{\sqrt{5}} + \dfrac{4}{\sqrt{5}} = 1 + \sqrt{5}.$$
Hence, we obtain that $M = 1 + \sqrt{5}$. Therefore $(M-1)^2 = 5$. $\square$

Problem 4.350. Let $p(x)$, $q(x)$, $r(x)$ be polynomials of degree 4 with real coefficients, such that for each real value of x it holds true $\sqrt{p(x)} + \sqrt{q(x)} = \sqrt{r(x)}$. Given that $p(1) = 0$, $q(2) = 0$, $r(1) = 4$, $r(2) = 1$. What is the value of $r(4)$?

(A) 1   (B) 16   (C) 100   (D) 289   (E) 400

Solution. Answer. (D)

Note that for any value of x we have that $p(x) \geq 0$ and $q(x) \geq 0$.
Therefore $p(x) = (x-1)^2 p_1(x)$ and $q(x) = (x-2)^2 q_1(x)$ for some quadratic trinomials $p_1(x)$ and $q_1(x)$, such that each of them is positive for any value of x. Let

$$p_1(x) = ax^2 + bx + c,$$

and

$$q_1(x) = mx^2 + nx + k.$$

We have that

$$\sqrt{p(x)q(x)} = \frac{r(x) - p(x) - q(x)}{2}.$$

On the other hand, we have that

$$\sqrt{p(x)q(x)} = |x-1| \cdot |x-2| \sqrt{p_1(x) q_1(x)}.$$

Note that

$$\lim_{x \to +\infty} \frac{\sqrt{p(x)q(x)}}{x^4} = \sqrt{a \cdot m} > 0.$$

Hence, we obtain that the degree of the polynomial

$$Q(x) = \frac{r(x) - p(x) - q(x)}{2}$$

is equal to 4. Moreover, we have that $Q(x) \geq 0$ for any value of x and

$$Q(1) = 0, Q(2) = 0.$$

Thus, it follows that

$$Q(x) = \sqrt{am}(x-1)^2(x-2)^2.$$

Hence, we deduce that, either

$$p_1(x) = a(x-1)^2, q_1(x) = m(x-2)^2,$$

or

$$p_1(x) = a(x-2)^2, q_1(x) = m(x-1)^2.$$

Therefore, either

$$r(x) = (\sqrt{a}(x-1)^2 + \sqrt{m}(x-2)^2)^2,$$

or

$$r(x) = (\sqrt{a} + \sqrt{m})((x-1)(x-2))^2.$$

Using the conditions $r(1) = 4$ and $r(2) = 1$, we obtain that

$$r(x) = (\sqrt{a}(x-1)^2 + \sqrt{m}(x-2)^2)^2,$$

and $m = 4$, $a = 1$. Hence, we deduce that

$$r(x) = ((x-1)^2 + 2(x-2)^2)^2.$$

Therefore $r(4) = 289$.   □

Bibliography

[1] MAA American Mathematics Competitions *AMC 12*

[2] Sedrakyan H., Sedrakyan N., *Number theory through exercises*, USA (2019)

[3] Sedrakyan H., Sedrakyan N., *How to prepare for math Olympiads*, USA (2019)

[4] Sedrakyan H., Sedrakyan N., *AMC 10 preparation book*, USA (2021)

[5] Sedrakyan H., Sedrakyan N., *AMC 8 preparation book*, USA (2021)

[6] Sedrakyan H., Sedrakyan N., *The Stair-Step Approach in Mathematics*, Springer Int. Publ., USA (2018)

[7] Sedrakyan H., Sedrakyan N., *Algebraic Inequalities*, Springer Int. Publ., USA (2018)

[8] Sedrakyan H., Sedrakyan N., *Geometric Inequalities. Methods of proving*, Springer Int. Publ., USA (2017)

Made in the USA
Columbia, SC
20 June 2023

18502519R00152